PUMPING IN THE CHEMICAL WORKS

THE CHEMICAL ENGINEERING LIBRARY.

ALREADY PUBLISHED:

THE GENERAL PRINCIPLES OF CHEMICAL ENGINEERING DESIGN

MATERIALS OF CONSTRUCTION: NON-METALS

WEIGHING AND MEASURING CHEMICAL SUBSTANCES

THE FLOW OF LIQUIDS IN PIPES

PUMPING IN THE CHEMICAL WORKS

IN PREPARATION:

THE LAY-OUT OF CHEMICAL WORKS

MATERIALS OF CONSTRUCTION: METALS

MECHANICAL HANDLING

FACTORY WASTES AS FUEL

Numerous other Volumes in preparation, 3s. net each.

PUMPING IN THE CHEMICAL WORKS

BY

NORMAN SWINDIN

WITH FORTY-FIVE ILLUSTRATIONS

LONDON: BENN BROTHERS, LIMITED
8 BOUVERIE STREET, E.C.4
1922

CONTENTS

PUMPING IN THE CHEMICAL WORKS

I

INTRODUCTION

PUMPING water, sewage, petroleum and non-corrosive liquids in quantity forms so large a part of ordinary mechanical engineering practice that all the requirements connected therewith may be said to be satisfactorily fulfilled. The pumping of corrosive liquids is an entirely different problem and now may be considered a distinct branch of engineering. To the chemical engineer the problem is mainly one of materials. If a suitable material can be found resistant to a certain corrosive liquid and capable of being fashioned into a mechanical pumping contrivance, ordinary mechanical engineering practice may be followed. However, many ordinary corrosive liquids attack all substances from which simple mechanical pumps can be made. The problem is in such cases of a different character and consists in applying energy to produce the movement of liquids without the intervention of mechanical aids—to remove the piston, the ram, the valve and springs, the glands and packing, the bearings and other weird things the mechanical mind has bequeathed to us.

It can be said that the problem of pumping corrosive liquids in the older chemical works was not even considered; it was in fact dodged, for such liquids were simply carried about in containers of fragile material. The problem, however, is compelling engineers to study first principles—the basic laws of fluid motion—and such success that the chemical engineer has achieved in the avoiding of mechanical devices is influencing the design of pumping plant in peculiar ways.

The development of the centrifugal pump is an attempt to cut out mechanical detail and to avoid the use of valves. The constant efforts to replace reciprocating motion by rotary motion in themselves simplify pumping mechanism because there is no reversal of inertia and therefore no need for fly wheels, dashpots, springs and dead centre stop devices, relief and safety-valves and air chambers. In this case the electric motor has had its influence on design. A high-speed motor geared down to run a ram pump is not used to the best advantage. Even the hydraulic ram pump is perhaps marked for the scrap heap and with it will go the accumulator.

As it is comparatively simple to construct containers for corrosive liquids of iron, steel, glass, lead, stoneware, ebonite and so forth, the older chemical manufacturers concentrated attention on the use of air pressure to force out the liquid through suitably arranged pipes, the valves and cocks being operated by hand. Such devices, known as acid eggs, are just about at the stage of development corresponding to the first Watt steam engine where the steam was admitted to the cylinders by valves operated by a man. Then came Platt, Kestner and others who fashioned automatic eggs by adding delicate mechanism. After these arose a greater one, Pohlé, who noticed that jets of air were somehow different from jets of steam and in a specification of remarkable clarity gave the first clear exposition of the air lift. The air lift even to-day is but a puny weakling, ill-developed and not understood, having about it no laws to explain its action and no rational formulæ to guide design. It is used generally where mechanical pumps cannot operate, as for instance the raising of gritty water from a deep bore-hole. Moreover, it is credited with low efficiencies, and in addition is debited with the losses of the air compressor. Under proper conditions the lift, considered as a pumping agent using energy of compressed air, has an efficiency of 80 per cent.

It will be shown later that when pumping hot liquids the air lift may have efficiencies apparently exceeding 100 per cent., in which case the heat in the liquid being pumped is converted into mechanical energy. Chemical liquids have to be handled at all temperatures up to their boiling points and they may contain in suspension precipitates, solids in various physical states, and their movement may be difficult. True, there are pumps advertised to pass solid bodies as large as clay bricks and other pumps with elastic insides which give and allow free passage for these bodies. The former are badly designed centrifugals, and the latter have efficiencies in some cases not so good as well-designed air lifts. The purpose of this work is to sketch briefly the general principles which will be useful to the chemical engineer in designing pumping plants. In the chemical works the problems encountered are of a more general character, and a knowledge of principles is of greater importance than in other locations, as not only may water have to be dealt with, but viscous, corrosive, gritty fluids at various temperatures may be involved.

II

CLASSIFICATION

The process of pumping consists in applying force to the end of a column of liquid in a pipe sufficient to cause movement. If the pipe is horizontal then movement simply means overcoming friction; if vertical, then the fluid is raised to a higher level in addition. The application of this force may be direct as in a ram pump and acid egg, or indirect as in centrifugal and air lift. Pumping appliances may be mechanical, semi-mechanical and non-mechanical. The theoretically-minded man would perhaps prefer the first classification, the so-called practical man the second. Both will be given to meet these views.

Classification No. 1.—The Manner of Applying Force.—*Direct-Acting Pump.*—(1) This includes all manner of ram pumps where energy from whatever source is first applied to a solid piston which pushes out or displaces from a cylinder a quantity of liquid previously charged. These again may be subdivided into those in which no provision is made for arresting and absorbing the momentum or inertia of the moving parts and those in which the ram or piston is brought to rest at the end of the stroke. In the latter class the Worthington pump is an example. Its development arose from endeavours to overcome the inertia difficulties in pumping petroleum over long distances in America.

(2) Septum displacement pumps are direct-acting, though the energy given to the piston is conveyed to the liquid to be moved through an intermediary device. This may consist of a rubber disc fixed so that the piston is protected from the corrosive action of the liquid, or a buffer of an inert liquid such as sulphuric acid or glycerine may be placed between the

piston or ram and the liquid being pumped. Examples of the former are the chemical works diaphragm pumps for hydrochloric acid and the rubber tube pump; of the latter chlorine pumps with a sulphuric acid buffer and the Ferraris pump.

(3) There are many types of rotary pumps: direct acting, such as the drum pump designed on the lines of a Roots blower, the multiple cylinder, the Douglas and the Brackett boiler-feed pumps, and in this class may be put a curious contraption known as the Frenier sand pump, which is a kind of air compressor and pump combined. It will be described later.

(4) It is legitimate to class as direct acting those lifting devices which depend on the pressure of a compressed gas being exerted in contact with the liquid. The most famous example is the Humphrey pump, where a combustible gas is exploded in a cylinder of which the water to be pumped may be considered as the piston. In the chemical works the time-honoured egg with its developments towards automatic working are examples.

(5) The fifth and last class consists of bucket elevators, the skin friction belt, the tank lift, the archimedean screw, and the rope and chain pumps.

Indirect-Acting Pumps.—In this class are placed those pumps in which the energy to cause flow arises from some change in the physical conditions or even composition of the liquid. Thus by whirling a liquid in a casing centrifugal force is generated in the mass of the fluid to cause flow through suitable passages into a pipe system. On this principle the centrifugal pump is constructed. Apart from the rotor there is an entire absence of controlling gear, valves, etc. The chemical engineer, so far as he can use a mechanical pump, adopts the centrifugal to the exclusion of ram pumps. The various forms of jet pumps, injectors, are further examples of pumping devices in which energy is conveyed indirectly by first imparting velocity to a fluid, steam or otherwise, which in turn transfers it to the liquid to be pumped.

Lastly, the air or gas lift in which a continuous stream of air or gas flowing into a column of a liquid causes motion as though the liquid were drawn up by a vacuum pump. The

Classification No. 1.

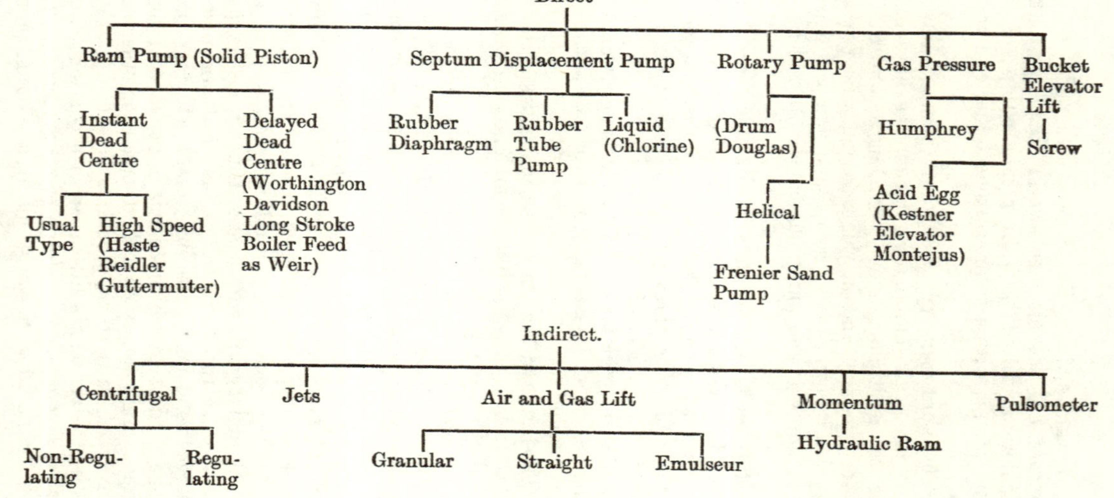

Fig. 1.

possibilities of this pump are not fully realized. Then there is the momentum pump, or hydraulic ram, where the momentum of a stream of liquid is absorbed in the momentum of a second stream when both flow out together. The lifting of mercury in electrolytic cells may be beautifully performed by this apparatus. The pulsometer acts indirectly, for it depends on the condensation of steam in twin chambers which are alternately opened and shut by automatic valves. It owes its popularity to absence of accessory gear, for it can be slung by a chain and requires little else but connecting up to a steam supply.

Fig. 1 shows this classification.

Fig. 2 shows the alternative system of classification.

CLASSIFICATION No. 2.

CONSTRUCTION OF PUMPS

MECHANICAL—

Ram Pumps of every kind — Instant Dead Centre. / Delayed " " / Valveless

Diaphragm or Septum Pumps.
Rotary Pumps.
Explosion Pumps (Humphrey).

SEMI-MECHANICAL—

Centrifugal Pumps of every kind.
Frenier Sand Pumps or Helical.
Pulsometer.
Tube Pumps.
Dancing Momentum Pump.
Kestner Elevator and Automatic Elevators.
Hydraulic Ram.

NON-MECHANICAL—

Air Lift.
Acid Egg, Montejus.
Jet Pumps.
Liquid Momentum Pumps.

FIG. 2.

III

THE RAM PUMP

The action of the ram pump is so obvious that its popularity is not surprising. Its action can be understood by the ordinary intelligent labourer, and it is credited with high efficiencies.

It is admitted that large, well-designed ram pumps have the highest efficiencies of any pumps, yet the efficiency of pumping operations in the ordinary chemical, tar, gas and similar works is generally not more than 10 per cent. The good points of the ram pump are that it is positive in its action; it does not "lose its water," and will generally work when in a very indifferent state of repair. No mysterious calculations are required to find the output or power required. Speed alone governs quantity, head is independent of both. The volume swept by the piston in a given time multiplied by a co-efficient K, assumed generally as 0·9, gives the quantity. The work done in lifting the water plus a bit for friction in pipes and pump give the horse power. It is all so very simple. Its disadvantages are legion. It possesses mechanism to corrode and to wear away, valve seats continually to be ground. Its lack of balance necessitates costly foundations. The repeated reversals of the flow of the liquid cause stresses which may shatter the pump and its connections. It is bulky and costly for a given duty, and often the extra interest on the capital cancels out any benefit from efficiency. Without the flywheel it is difficult to drive by belt, steam or gas engine or motor. Ram pumps cannot handle easily dirty, gritty and highly viscous solutions.

The action of the ram pump is not, however, so simple as it looks. Because of the incompressibility of liquids and their great densities compared with gases, the reversal of flow and sudden changes of velocity produce effects which are not yet understood.

Prof. John Goodman in 1903 made some careful experiments on the behaviour of an ordinary ram pump. Aware that Henry L. Worthington had designed a direct-acting free piston pump which successfully obviated the frequent bursts in long pipe lines of the American petroleum industry, and of the failure of certain locomotive boiler feed pumps, he set out to study the conditions which produced the shock pressures which lay at the root of these troubles. He first fitted a special hydraulic indicator to the delivery pipe of one of the boiler feed pumps, and found that the maximum pressure recorded was 3,250 lb. per sq. in., though the pump was only pumping against a pressure of 140 lb. per sq. in. On increasing the size of the passages and otherwise easing the flow of water, this pressure was reduced to 850 lb. per sq. in. Theory could only account for 200 lb. in the first case and 150 lb. in the second. It was discovered that the water separated from the plunger at a speed of about one-third of that at which the pump was running.

The problem stated is to determine:

(1) How the "slip" or the "discharge co-efficient" of a pump not fitted with a vacuum vessel, also the "water ram" pressure in the suction pipe are affected by: (*a*) A change of outlet or delivery pressure when the speed remains constant. (*b*) A change of speed when the delivery pressure remains constant. (*c*) A change in the length of the suction pipe with the other conditions remaining unchanged. (*d*) Running the pump without a suction valve.

(2) The exact behaviour as regards the opening and closing of the suction and delivery valves under various conditions of running.

(3) The speed at which the plunger separates from the water during the early part of the stroke and catches it up later on, thereby producing a shock in the suction pipe.

(4) The loss of pressure due to the friction of the water passing through the valves and passages of the pump.

(5) The mechanical efficiency of the pump under various working pressures.

(6) The effect of fitting a vacuum air vessel to the suction pipe.

By a discharge co-efficient is meant:

$$\frac{\text{Vol. of water actually pumped}}{\text{Vol. displaced by the plunger}} = S.$$

The results of Goodman's experiment showed that when the pump is running smoothly it is possible to pump more liquid than corresponds to the displacement of the piston, i.e. that slip is negative.

Engineers are accustomed to the use of a "discharge co-efficient" value of 0·9 to 0·95. It may surprise some of them that it is possible to pump 1·5 times more liquid than the volume swept out by the piston. When a ram is working very slowly the diagram taken by the indicator is practically a rectangle, thus (Fig. 3):

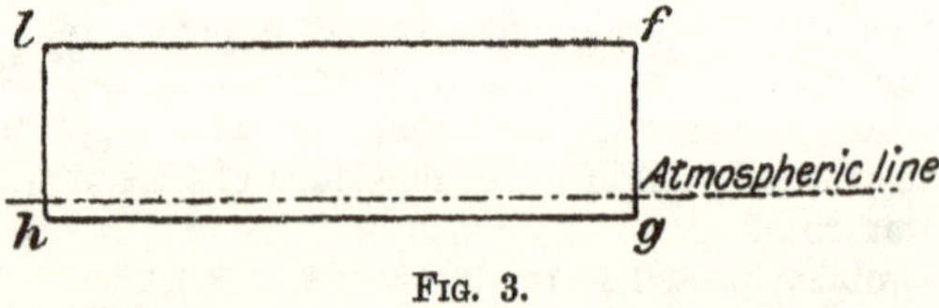

Fig. 3.

If now the speed of the ram is increased the inertia of the moving mass of water affects the pressures, and the diagram changes to something like Fig. 4.

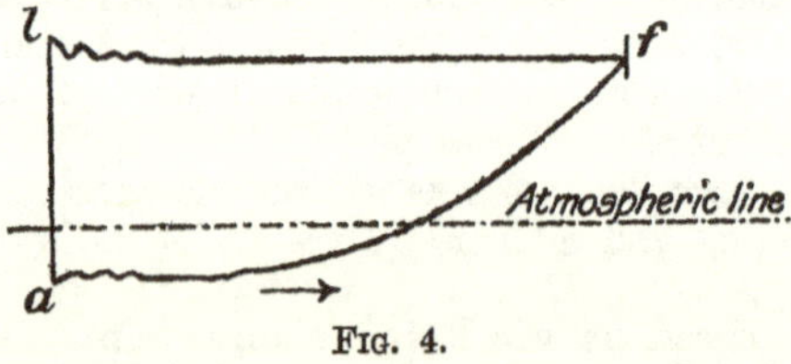

Fig. 4.

The suction line *af* is no longer a horizontal line, but owing to the retardation of the ram towards the end of stroke at point *f* the momentum of the water in the suction pipe is arrested, producing sharp rise of pressure. By further increasing the speed of the pump the momentum of the moving mass of liquid in the suction pipe is large enough to lift the delivery valve before the ram has reached the end of its stroke. Fig. 5 gives an actual diagram taken from Prof. Goodman's paper showing this anticipated lifting of the valve.

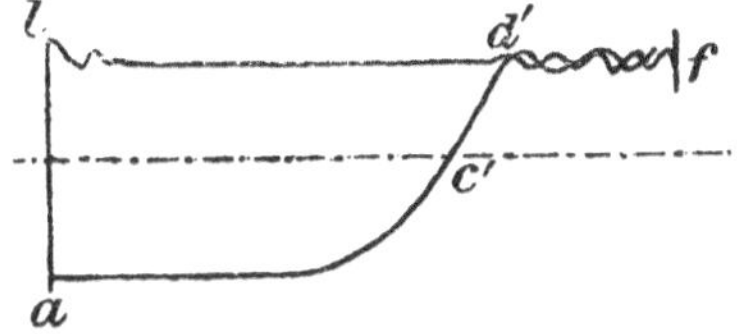

FIG. 5.

In such cases the discharge of a greater volume of liquid than the volume displaced by the ram is evident. Further, it is possible to pump without the use of suction valves, the column of liquid in the suction pipe itself acting as a sort of valve. This action of a ram pump, on account of the range of liquids to be moved, is of great importance to chemical engineers, so that it will be considered in greater detail. Let Fig. 6 represent the theoretical diagram of a pump working under such conditions, but neglecting friction for the present.

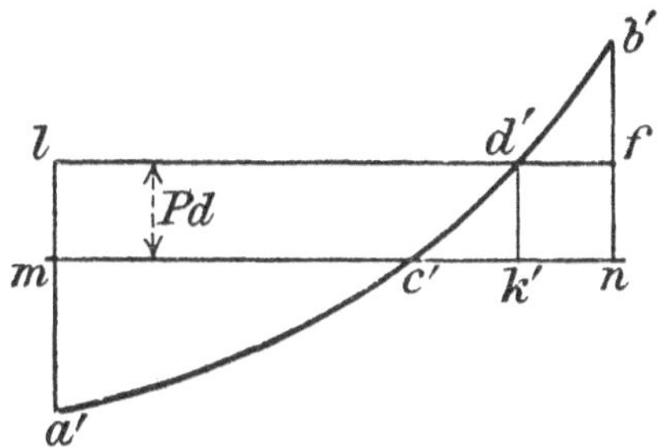

FIG. 6.—Theoretical Diagram.

The line $a'b'$ represents the distribution of pressure due to the inertia at all parts of the stroke. From a' to c' the pressure is negative because the pump is accelerating the water in the suction pipe. At c' the inertia pressure becomes zero and in consequence of the retardation of the motion of the liquid a pressure is actually exerted on the plunger. At d' the inertia pressure equals the delivery pressure, causing the liquid to open the delivery valve and pass away before completion of the suction stroke. Since the ordinates of the curve $a'b'$ represent the water pressure on the plunger, and the abscissæ the horizontal distances through which the plunger has moved, the area $c'd'k'$ represents the work done by the retarded water in forcing the plunger forward, and the area $d'b'f$ represents the work done in delivering the extra water during the suction stroke. Now work done is the product of volume of liquid and the pressure so to the extra volume of liquid pumped is area $d'b'f$ divided by the delivery pressure. Let this extra volume be v and let V denote displacement volume of plunger, then the discharge co-efficient: $S = 1 + \frac{v}{V}$. Since the ratio of the work done in delivering the extra volume of water to the work done under normal conditions is $\frac{\text{Area } d'b'f'}{\text{Area } lfmn}$ and since the pressure scale is the same in both cases, the volume of water delivered will also be in the same for operation, whence the discharge co-efficient:

$$S = 1 + \frac{\text{Area } d'b'f}{\text{Area } lfmn}$$

In order to calculate the inertia pressure for a pump having an infinitely long connecting-rod, let:

R = the radius of the crank in feet.

L = the length of connecting rod in feet.

$n = \frac{L}{R}$

N = revs. of pump per min.

w = weight of unit column of water one foot high and one sq. in. section = 0·434 lb.

W = weight of reciprocating parts per sq. in. of plunger in lb.; in this case weight of column of water 1 sq. in. section and length = to suction main, i.e. 0·434 L = wL.

P = the "inertia pressure" at end of stroke, i.e. the pressure required to accelerate and retard the column of water at the beginning and end of stroke.

Then if the suction-pipe be of the same diameter as the pump plunger we have: $P = 0{\cdot}00034\ WRN^2$, but if the area of the suction pipe be A_s and that of the plunger A_p we have substituting the value of W given above:

$$P = 0{\cdot}00034 \times 0{\cdot}434\ LRN^2\ \frac{A_p}{A_s}$$

The manner in which the pressure varies from point to point in the stroke is given by the straight line *ab* (Fig. 7), and the

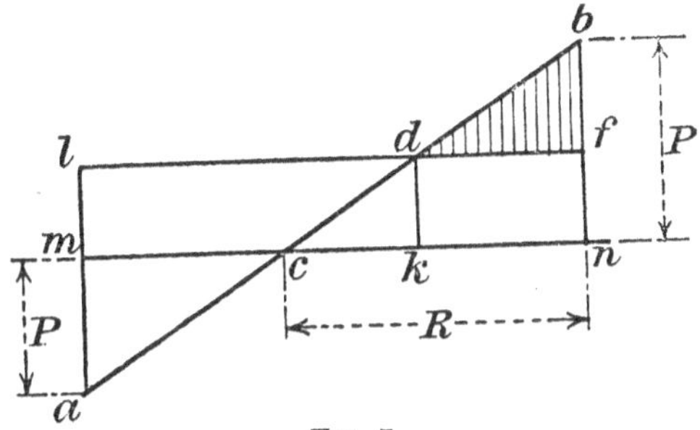

FIG. 7.

extra volume of water delivered is equal to the area $\frac{dbf}{144P_d}$ calculated to the appropriate scale of the diagram.

For a pump having a short connecting rod

$$P = 0{\cdot}00034\ WRN^2 \left(1 \mp \frac{1}{n}\right)\frac{A_p}{A_s}$$

The line *ab* (Fig. 7) becomes the curved line *a′b′* (Fig. 6).

What happens when the ram is moving at high velocities is not even to-day clearly perceived. For every ram pump there is a critical speed—the speed at which the piston moves faster than the water entering the cylinder. Under these conditions the piston leaves the incoming water, and later on in the stroke, when the speed of the ram pump or plunger is reduced, the water will catch it up and cause a bang when they meet, the violence of the bang depending on the velocity of impact. The speed at which the separation of the plunger and water occurs can be calculated.

Speed at which Plunger leaves the water in the Pump Barrel when not fitted with a Vacuum Vessel.

Let h_s = the suction head before the pump, i.e. the height of the surface water in the sump below the bottom of the pump barrel; if it be above, this quantity must be given the negative sign.

h_f = the loss of head due to friction in the passages and pipes.

Then the pressure required to accelerate the moving water at the beginning of the suction stroke is

$$P = 0{\cdot}00034 \times 0{\cdot}434 \ \text{LRN}^2 \left(1 - \frac{1}{n}\right) \frac{A_p}{A_s}$$

(Notation as on page 18.)

If the barometer be taken as 34 ft., then the effective pressure driving the water into the pump barrel is $(34 - h_s - h_f)$.

Separation occurs when this quantity is less than P; equating these two quantities we get the maximum speed, N, at which the pump can run without separation taking place, and reducing we get

$$N = 54{\cdot}5 \sqrt{\frac{(34 - h_s - h_f)A_s}{\text{LR}\left(1 - \frac{1}{n}\right) A_p}}$$

In cases in which friction plays an important part, the separation will not necessarily occur at the beginning of the stroke because the water is then at rest and the friction zero, but the frictional resistance increases rapidly as the plunger moves: hence the separation will occur shortly after the beginning of the stroke.

At this critical speed the discharge co-efficient is very unstable. Goodman's experiments on his small pump give results from which Fig. 8 has been plotted.

Chemical works ram pumps fall into two main classes; the first includes those which pump non-corrosive liquors, and the second those which pump corrosive liquors. The former are constructed of ordinary cast iron, steel or bronze, while the latter must be made of special materials to resist corrosion.

Of the non-corrosive liquors little here need be said concerning the best type of pump to raise water, this subject being fully treated in works on mechanical engineering. The non-corrosive liquids of the chemical work are volatile limpid liquids like toluol or benzol, or heavy viscous liquids like coal tar, glycerine, soap solutions, oils, melted fats and waxes, magmas such as paper works "stuff" and slurries, like milk of lime. Any ordinary ram pump may be used for the limpid liquids, but pumps for the moving of tar, glycerine and similar substances must be designed on ample lines. The passages in

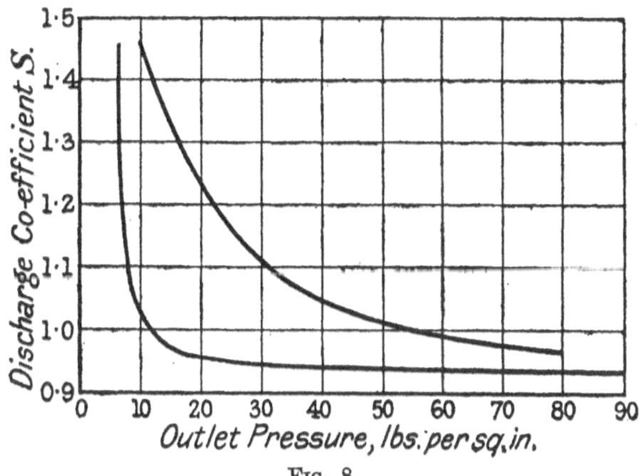

Fig. 8.

and out of the cylinder and the valves must be made larger than usual, and it may be necessary to fit steam jackets, or at least steam pipes, to the pumps in order to reduce the viscosity of these heavy solutions. The speed of running also is much lower than that adopted for pumping water. There are pumps on the market in which the valves are mechanically operated and take the form of a piston or slide valve worked from an eccentric.

The Duplex pump first introduced by Worthington works satisfactorily where viscous solutions offer great frictional resistance. The momentary stop of the ram at the end of the stroke allows the inertia to be absorbed and dispenses with the use of air vessels on either suction or delivery side. This point has already been discussed at the opening of this chapter.

Gritty substances had better be kept out of a ram pump, but if it is not possible then the loosely fitting ram with soft packing must be adopted. The well-known Cameron pump is an example of this type. It is an advantage to fit a cup on the gland in order to retain a small quantity of oil or even water.

Speaking generally, the pumping of viscous liquids is best done by some form of rotary pump to be described later.

The pumping of corrosive substances by a ram pump is a problem chiefly of materials and stuffing box design. Earthenware pumps are obtainable, but are only adopted where pumps of more sturdy material cannot possibly be used. Great care is required to keep the glands tight, otherwise the leaking acids corrode the metal clips and fastenings, which hold the earthenware parts together. For hydrochloric acid ebonite pumps are superior to the earthenware type, but almost the same care is demanded. Ebonite is brittle and liable to snap, especially so if any attempt has been made to cheapen the product by the addition of ground-up vulcanized rubber, which merely acts as a filler. For pumping hot solutions ebonite pumps cannot be used, as softening takes place at about 100° C., with the certain consequence of seizing. Further, ebonite has a tendency to harden if used continually for even warm solutions. The type usually seen is rather a cumbersome piece of work and designed on the line of a hydraulic pump. The plunger, which is double-acting, is kept tight by hydraulic grooves instead of packing. It is now possible to obtain ebonite-lined centrifugals and diaphragm pumps and acid eggs, which on the whole are to be preferred to the plunger type of pump.

Ram pumps of ferro-silicon alloys are not recommended. This material is so difficult to machine and cast that it is better to adopt the centrifugal design of pump. These alloys resist nitric and strong sulphuric acid perfectly, but they must not be used for hydrochloric, sulphurous, very weak sulphuric, solutions containing ferric chloride, and organic acids of the type of malic and tartaric.

Kestner claims to have overcome the packing difficulty. In his pump the ram is made extra long and the stuffing box is replaced by a cylinder with only a working fit. A film of acid is really the packing and the slight leakage, which amounts to only about two drops per minute, is returned to the suction.

IV

THE SEPTUM DISPLACEMENT PUMPS

THIS type of pump represents the first serious efforts to remove the propelling mechanism from the action of corrosives. The ram or plunger chamber is separated from the pump proper by a diaphragm of rubber, or a buffer of an inert substance, such as oil, is placed between the ram and the liquid being pumped. The illustration, Fig. 9, shows the ordinary form of rubber diaphragm pump. A. L. G. Dehne, Halle, supplies a satisfactory type of this pump, but now several English makes are available. The plunger works in water, the displacement of which deflects the rubber diaphragm from side to side of a flattened chamber. The acid side of the apparatus comprises the usual valve boxes and air vessels lined with suitable material. The valves are fitted with rubber balls. These pumps should not be used for hot liquors, but they are well adapted for pumping slimes, precipitates and gritty substances—in fact the design owes its development to the use of the filter press.

The diaphragm pump is now largely adopted in mines for pumping slimes for low lifts (25 ft.), and when fitted with adjustable stroke it is superseding the air lift. It is well suited for handling the sludge from mechanical thickeners of the Dorr type.

There is an old form of septum pump which operates by the alternate squeezing and releasing of a rubber tube, and is still in use in certain mining districts. Usually it consists of a rubber tube about 3 in. diameter, the lower end of which is surrounded with a casing with two swellings each communicating to a cylinder fitted with a plunger. The space enclosed by the swellings and the plunger is filled with water. The plungers are geared so that alternate squeezing and releasing of the rubber in the swelling takes place and so pumping the liquid.

A variant of this rubber tube septum is now on the market in which two rollers attached to the extremities of rotating arms in a suitable casing squeeze a rubber tube lying in a

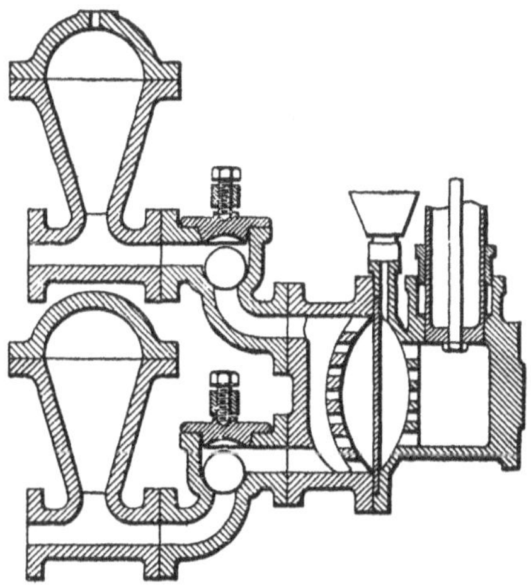

Fig. 9.—Diaphragm Pump.

slightly eccentric path in this casing, and some hydrochloric acid pumps are made on this principle.

The Ferraris acid pump is a liquid septum pump for sulphuric and hydrochloric acid. A column of suitable oil acts as the buffer, and the limit to its usefulness is the choice of oils. The pump is of Italian origin, but has not found extended use in this country. Where first cost is not important it is one of the most reliable pumps for sulphuric, being automatic compared with the egg, and not requiring submergence wells when compared with the air lift.

V

THE CENTRIFUGAL PUMP

If there were a material as resistant to acid corrosion as iron or steel is to water then the centrifugal pump would make the wildest dreams of a chemical engineer come true. Materials apart, the centrifugal has nearly all the qualities of a chemical pump. Compared with a ram pump it is simplicity itself. It has no valves, springs, and no mechanism: instead an impeller mounted on a simple shaft. The ordinary engineer usually insists on the gland and stuffing box, but the chemical engineer has even abolished this. The centrifugal delivers a steady stream of liquid, which no ram pump can possibly do. It delivers anything from 20 to 100 times the quantity of liquid as a ram pump of the same size. Originally developed as a low lift pump, the centrifugal has now ousted the ram pump from its special domain, the high pressure hydraulic service. Pressures of 800 lb. per sq. in. can be obtained by compounding, and Rateau has designed a single impeller to pump against 863 ft. head.

The centrifugal will pump mud, sand, stones, and lumps of anything if these things can enter at all. Here the ram pump downs tools at once and goes on strike for ever. With the centrifugal the dredging engineer scoops up the river bed, and a form of this pump exists which has flexible blades in its impeller in case an extra large object enters.

The design of the centrifugal is an expert's job, and the reader must consult him and study his works for a complete analysis and account of its action, but the installing and working of this pump calls for more than the average chemical foreman can give.

The writer has many sad experiences to relate concerning

the installing of centrifugal pumps. One of the largest works pumping stations ever erected was almost stifled at birth by the intrigues of a class of foreman whose knowledge of engineering did not extend beyond the fitters' bench. Let it be understood that no one who has not grasped the full significance of $V^2 = 2gH$ should be allowed to look at a centrifugal pump, much less install one.

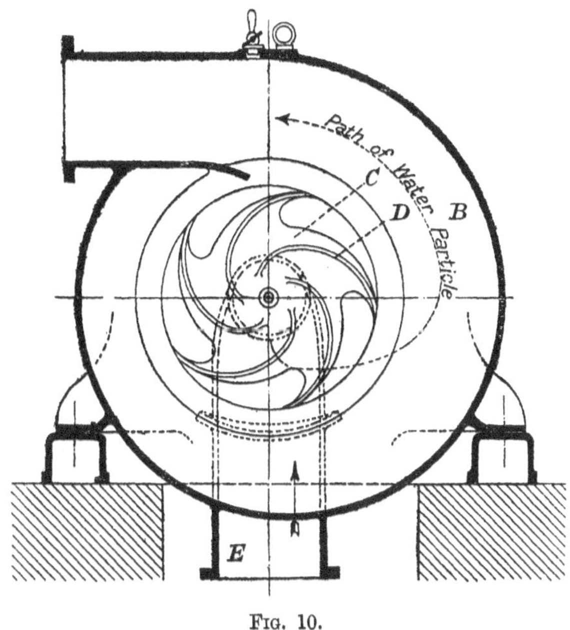

Fig. 10.

Fig. 10 shows in diagram the centrifugal pump. The liquid entering the pump at E is rotated at a high speed by the blades D of the impeller C, and flung off the periphery of the impeller at a velocity which is the resultant velocity of the tip of the blade of the impeller and the radial flow of liquid through the pump. This velocity of the water is gradually arrested and the energy of motion converted into pressure head in a suitable casing B in which the impeller rotates. The liquid is now under pressure and will flow out of the pump against the resistance to be overcome. The design of the pump is directed towards the conversion of the energy of velocity to potential energy without loss, minimizing the friction of disc rotation and balancing the end thrust of the impeller shaft.

At this point it is as well to state very briefly the principles upon which the centrifugal pump works, and to describe the

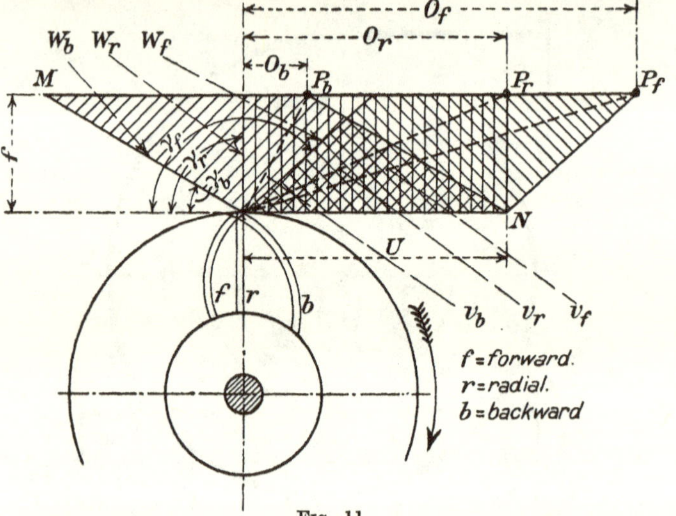

FIG. 11.

various methods in which these principles are applied in practice.

Let Fig. 11 represent a centrifugal impeller having three types of blades—radial *r*, backward curved *b*, and forward curved *f*.

The liquid enters the eye of the impeller shown by small circle, Fig. 11, and for simplicity's sake let us assume without shock. As the impeller rotates in the direction indicated by the arrow, the liquid is whirled round and slides along the surface of the blade and is flung off at a velocity V. Now V is a component of two other velocities; one, U, that of the impeller tip which is $\frac{ND}{60}$ ft. per sec. where:

N is revs. per min.

D is dia. of impeller in feet

in a direction tangential to the periphery—the other, W, tangential to the curvatures of the blade itself and magnitude represented by the intercept of the radial velocity f of the liquid through the pump. To draw the parallelogram of velocity at the tip of an impeller blade, proceed thus:—

Draw U tangent to periphery at tip of blade and length representing velocity of the tip.

Parallel to U draw a second line distance f the radial velocity of the water through pump.

If Q = cu. ft. liquid passing per sec.

A = area of impeller discharge opening = D × K, where K is width of blade in feet,

then $f = \frac{Q}{A}$

From tip of blade draw tangent W intersecting parallel line at M.

From point N on line U draw NP parallel to W.

Lastly join P to tip of blade.

Then parallelogram M P N tip is velocity diagram, in which U is velocity of impeller tip; W velocity of water along blade surface; V velocity of liquid leaving impeller, f radial velocity of liquid through pump.

The aim of the designer is to arrange the blades so that the friction of the liquid along them shall be a minimum and that the velocity of the liquid V leaving the impeller shall also be minimum. The equating of these two factors is a prime consideration in design.

Referring again to Fig. 11, it is seen that the more the blade is curved backward the less is the value of V, but also the greater becomes W. The value of V governs the loss in converting energy of whirl into pressure head, i.e. the less it is the less is the loss, but W is a function of friction loss, so that the greater W the greater is the blade friction.

Notwithstanding the tomes of theory written on this subject experience here is the chief guide, but the following may be accepted as a rough rule:—Backward curved blades for high lifts, radial for low and medium lifts, and forward blades for very low lifts such as those fitted in irrigation pumps, where water is only raised a few feet. An authoritative set of experiments to determine the efficiencies of radial and curved impeller blades were published by Dr. Stanton in 1903. From these experiments it was concluded that in high speed wheels, i.e. wheels in which the velocity of the tips of the vanes exceeds 40 ft. per sec., the effect of moderately recurving the vanes at the outlet is beneficial, the velocity of flow of water through the wheel is uniform.

These results are confirmed by Mr. J. A. Smith, of Melbourne,

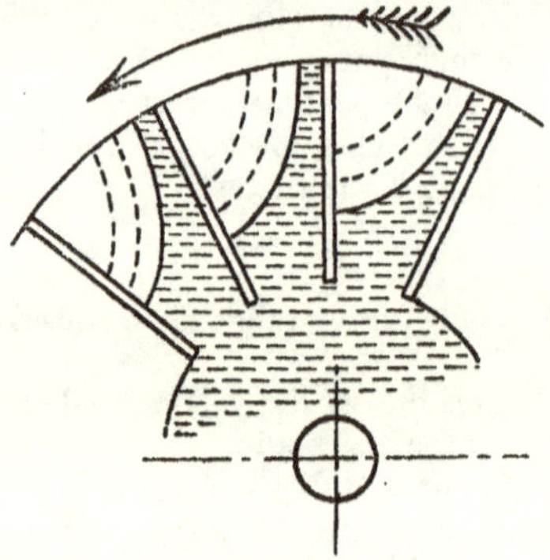

Fig. 12.

who took instantaneous photos of the flow of water through an impeller discharging freely into the air. A study of these photos is of interest to the chemical engineer because of the importance of viscosity and density, which the hydraulic engineer does not worry about. All the theory of the centrifugal is based on liquids composed of free moving particles. Even Gibson in his *Hydraulics and its Applications* states in his usual complete way the factors governing the design of the pump admits that " none of the fundamental assumptions made are true . . . since viscosity must cause a defection of the lines of flow in the direction of rotation." Smith's results, Fig. 12, show the liquid leaving the leading edge of the blade and forming a layer of dead liquid shown white. This dead water is in a state of eddy and therefore a source of loss.

Work done on Pump $= \frac{WQ}{g}$ (Or) where W= weight of cu. ft. liquid

Turning effort on shaft.

Q = cu. ft. per sec.
r = radius of impeller tip
g = 32·2 ft. per sec.
w = velocity radians per sec.
O = } Velocities, *see*
U = } Fig. 11

$\therefore$ Work done on water per sec. $= \frac{WQ}{g}$ (Or) w ft. lb.

$$= \frac{WQ}{g} \text{ OU}$$

$$= \frac{\text{OU}}{g} \text{ ft. lb. per lb. liquid.}$$

This does not include for losses in eddy formation, shock, friction of liquid flow, mechanical friction and leakage.

Energy from Pump :

Referring to Fig. 13.

If H is difference of level between suction and discharge reservoir,

H_f friction loss in suction and delivery pipes,

v velocity of flow along discharge pipes,

then energy from pump per lb. is :

$$H + H_f + \frac{v^2}{2g} \text{ ft. lb.}$$

Theoretical Manometer Efficiency

$$= \text{ratio } \frac{H_m \text{ per lb. water}}{\frac{\text{OU}}{g}} = \frac{H_m g}{\text{OU}} = \frac{H_m g}{U\,(U - f \cot.\ \gamma)}$$

Hydraulic Efficiency

$$= \eta_h = \frac{H_m}{\frac{\text{OU}}{g} + L_h}$$

where L_h = total loss of energy in overcoming hydraulic resistance in pump itself.

Working Efficiency $= \dfrac{\text{Energy from pump}}{\text{Energy given to pump.}}$

If $N =$ revs. per sec.
$Q =$ cu. ft. per sec.
$T =$ turning moment in ft. lb.
then:

$$\eta = \frac{QWH_m}{2\pi NT} = \frac{H_m}{\dfrac{OU}{g} + \dfrac{L_h + L_m}{WQ}}$$

where L_h = hydraulic losses.
L_m = mechanical losses.

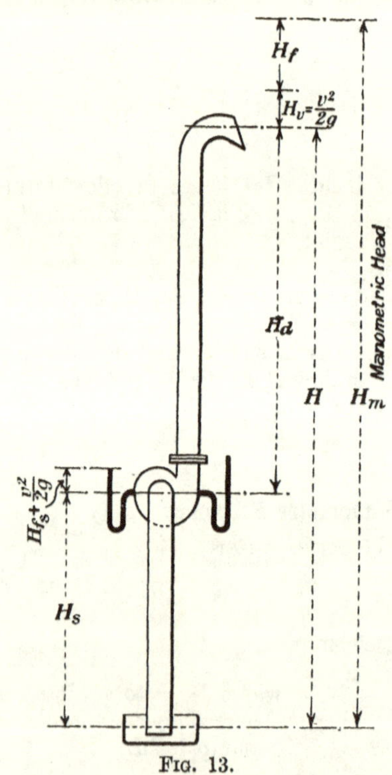

FIG. 13.

As pumping may be defined as the art of exerting pressure on a column of liquid, centrifugal pumps may be classed according to methods of converting velocity head to pressure. This conversion is one of some difficulty, and a variety of centrifugal pumps have been evolved to carry this out with the highest efficiency.

In the simplest form of centrifugal the liquid is allowed to discharge from the periphery of the impeller into a circular chamber or casing, no special provision being made to prevent shock, as in Fig. 14.

A slight improvement on this is the replacing of the circular chamber with a volute casing, thus (Fig. 15).

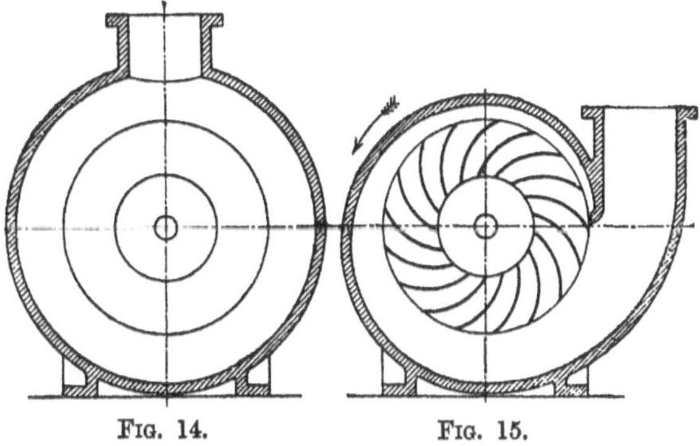

FIG. 14. FIG. 15.

There the liquid discharges freely into a volute chamber in which the velocity of whirl is uniform. There is, however, still a loss by shock because the velocity of the liquid leaving the impeller is greater than the velocity in the involute chamber. The gain in efficiency due to this chamber is, according to Stanton's researches, only about 10 per cent.

Next in development is the pump with whirlpool or vortex chamber, a device first adopted by Prof. James Thompson. In this pump the impeller is surrounded by a circular chamber upon which is superposed the involute casing as in Fig. 10 above. This circular chamber increases the dimensions of the pump and consequently the cost. In practice a compromise is adopted, both involute and whirlpool chambers are cut down.

Owing to the tendency for the liquid to form eddies, notwithstanding all the previous devices to convert velocity energy into pressure gradually, a system of guide vanes fixed round the impeller is now the recognized practice in the best type of pump. These vanes receive the water from the impeller without shock and direct it by gradually diverging passages to the whirlpool or involute chambers. A centrifugal so fitted is like a reversed water turbine, and is aptly termed a turbine pump (*see* Figs. 16 and 17). When properly designed 75 per cent. of the kinetic energy of discharge is converted into pressure.

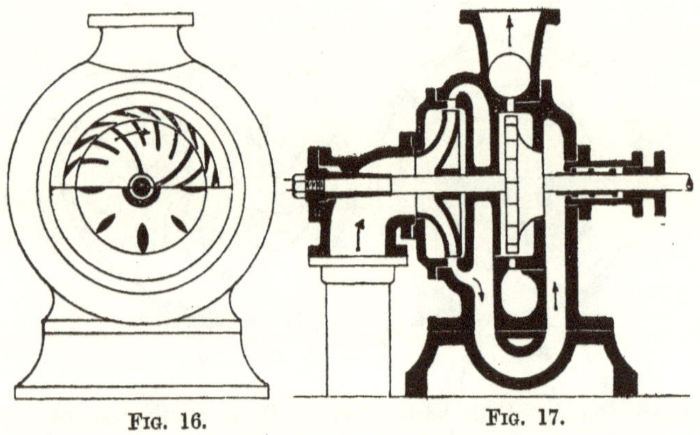

Fig. 16. Fig. 17.

There are modern types in which the conversion takes place entirely within the impeller. One such pump has an impeller surrounded by an annular Venturi throat guide vane, vortices being entirely avoided, while another replaces the flat impeller by a revolving pressure drum.

The design of centrifugal pumps for moving chemical liquids is determined chiefly by the limited choice of materials. Manufacturers therefore have wisely adopted simple forms of the pump at the expense of efficiency. The following description of some well known makes is given simply to illustrate guiding principles, and to enable the reader to form his own judgment on the merits of any particular pump.

Where cast iron, steel or bronze may be employed standard practice may be followed and any of the well known makers of centrifugal pumps may be trusted to supply a suitable pump if given a full statement of the conditions under which it shall work. This statement should include the quantity of the liquid to be moved, the nature of the liquid, its specific gravity, viscosity, whether gritty or volatile, the suction and delivery heads.

The selection of a centrifugal pump is governed by :—

1. The quantity. This should be stated in gallons per minute, together with the limits of variation and normal quantity. A glance at the characteristic curves of an ordinary medium lift pump shows that the quantity cannot be altered without affecting the head ; the head falls as the quantity is increased.

2. The nature and properties of the liquid ; i.e. the density, for this determines the head ; the variation of density and viscosity with regard to temperature. The failure of many a centrifugal is due to a sudden increase in resistance or head on account of the lowering of the temperature. The resistance to flow of saturated brine is $1\frac{1}{2}$ times as great at 10° C. as it is at 90°C. The ratio $\frac{\text{viscosity}}{\text{density}}$ is a measure of the resistance. Under this head should be stated the volatility of vapour pressure for a temperature range, for this determines the suction head, for a centrifugal will not pump if air or gases are given off in the casing (whenever possible the pump should be arranged with a small pressure head at the inlet) ; also state whether gritty or not (grit wears the neck rings and diminishes the efficiency).

3. The delivery head :—The delivery head determines the speed of the pump, and it must be accurately determined, and given either in lb. per sq. in. or in feet of the liquid of a given density. In many cases it is exceedingly difficult to ascertain, for frictional resistances of many chemical liquids are unknown. The Stanton Curve, as given in *Flow of Chemical Liquids*, in this series, should be consulted and adequate values for the resistance of valves, bends, etc., obtained.

4. The suction conditions. Whether the liquid flows to the

pump by gravity or has to be lifted ; if it has to be lifted state lift in feet and limit of variations calculated from pump centre; also length of suction pipe and number of bends, noting beside the temperature and vapour tension of the liquid.

To start a centrifugal with suction lift, special priming arrangements are necessary, sometimes an injector, pump or a simple cock and funnel, together with foot valve.

For chemicals liquids especially try to arrange suction under a slight head, even at the expense of a vertical driven pump.

Some Chemical Centrifugal Pumps.—Though Messrs. Mather and Platt, Ltd., Manchester, do not make the special type of small centrifugal for chemical works this firm deserves mention as being pioneers in the development of the modern centrifugal and turbine pump. For water, boiler feeding and oil where fair volumes are to be handled, this firm's pumps are to be relied on, and where the conditions are definitely stated guarantees are given. These pumps are usually electrically driven; as the motor is fully protected against overload no special self-regulating devices in the pump itself are adopted.

The Rees Roturbo and the Sunturbo are centrifugals of the self-regulating type. The rotor of the Rees Roturbo takes the form of a pressure drum, the object being to obtain by centrifugal force inside the drum a constant hydraulic pressure. In the rim of the drum a series of turbine nozzles are formed which are designed to impart to the water a resultant velocity one-half that of the peripheral velocity of the drum. It is claimed that only 25 per cent. of the total energy is converted into velocity, the water leaving the impeller at 25 per cent. velocity and 75 per cent. pressure, thus simplifying the design of the guide blades and passages, and greatly reducing wear and tear. This firm have recently put on the market a line of pumps specially designed for corrosive liquors, and constructed of special metals. In addition to the ordinary gland packed horizontal pump, Messrs. Rees Roturbo supply a vertical spindle glandless type.

The " Sunturbo " pump also has self-regulating properties due to the design and construction of the " Sunturbo " Patent Venturi impeller. This impeller follows the Venturi Law, giving a high velocity of flow at the blade tip or throat and then converting the velocity created into pressure before the liquid leaves the impeller. As the velocity of flow increases the pressure diminishes, and at the point where the liquid leaves the blades of the impeller a partial vacuum is created reducing wear and corrosion to a minimum and ensuring a high and permanent efficiency. Consequently no special volute chambers or guide blades are required; the impeller is fitted accordingly in a simple circular casing as in Fig. 14. The degree of self-regulation is seen from the diagram Fig. 18. Messrs. Meldrums, of Timperley, supply this type of pump in a ferro-silicon alloy for chemical works.

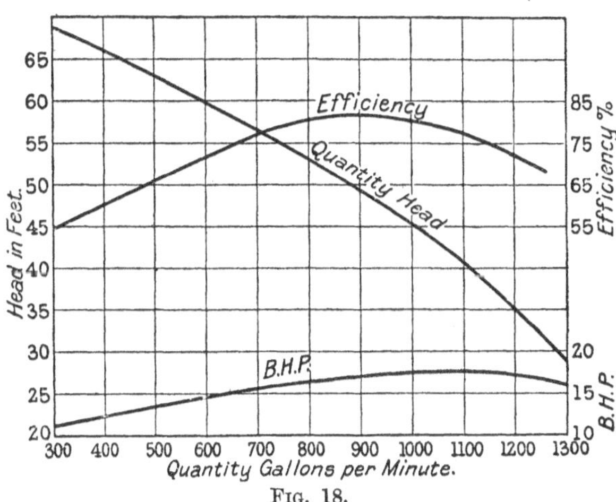

FIG. 18.

Centrifugal pumps of ferro-silicon alloys are now made by many firms. In England there are the Lennox Foundry Co., Ltd., London (Tantiron), John Varley & Co., Ltd., St. Helens (Narki), Haughton's Patent Metallic Packing Co., Ltd., London (Ironac), Meldrums of Timperley, and probably others, for the manufacture of these alloys is not now confined to one firm. In U.S.A. there are Bethlehem Steel Co. (Corrosiron), Duriron Castings Co., Dayton, Ohio (Duriron). These small pumps are gradually displacing the acid egg because of their automatic action, small cost and size. The design centres round the foundry and grinding shop, for these alloys are exceedingly hard and somewhat brittle, The casting of an impeller is almost a work of art, and all the machined surfaces must be accessible to the grinding wheel. The points to watch in deciding upon the purchase of a ferro-silicon pump are accessi-

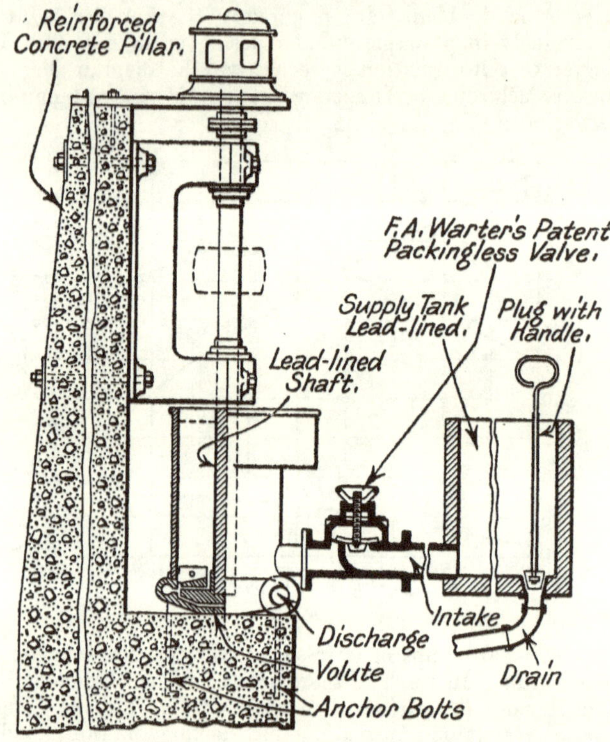

Fig. 19.

bility without disturbing pipe lines, positive and automatic lubrications of the shaft, the stuffing box preferably being liquid sealed and fixed always on the suction side. Some designs, such as the Duriron pump, permit of the suction and delivery adjustable to any position. The fixing of the special metal parts to the cast-iron bed plate is important—small feet and brackets are liable to crack and break off and special care is to be taken to see that the slotted flanges are properly ribbed and strengthened. Fig. 19 shows a glandless centrifugal acid pump by the Chemical Pump and Valve Co., N. J., U.S.A.

Ferro-silicon alloys make excellent liners for centrifugal pumps designed for pumping hard and gritty substances. Tailings pumps, dredge pumps and pumps for coal and coke washing plants are now obtainable with such linings. When required to pump acids to any considerable head, say 60 feet, provision must be made for taking up the end thrust. This is usually done by mounting some of the standard types of thrust ball bearings. It must be remembered also that ferro-silicon impellers cannot be rotated at speeds which are quite safe with steel or bronze impellers, so that to pump acid against heads exceeding 60 feet it is better to adopt the stage pump. A well designed line of stage pumps in Ironac is now on the market. The following table, supplied by the makers, gives the duties of three sizes of pump in two, three or four stages for pumping sulphuric acid of 1·5 sp. gr. (100° Tw.). The efficiencies of the 1½ in. and 2 in. are from 40 per cent. to 48 per cent. and from 55 per cent. to 62 per cent. for the 3 in. and 4 in. size.

Makers are urged to present the duties of their design of centrifugals in the standard form of characteristic curves similar to those shown in Fig. 18. It is noticeable how timid are chemical pump makers in publishing the results of carefully made tests. Many of the tests submitted for this publication are not reliable and most have been carried out in a rough and ready manner. On the average it may be taken that this type of pump possesses working efficiencies of 30 per cent. for 1 in. pump, 40 per cent. for 1½ in., and 45 per cent. for 2 in. Larger sizes have efficiencies 20 per cent. lower than the ordinary water centrifugal. The ultimate form, in the writer's opinion, of chemical works centrifugal will be the freely suspended, vertical and glandless type.

Regulus Metal Pumps.—For ordinary sulphuric acid free from nitro bodies the pump consisting of a regulus metal casing and ferro-silicon impeller is recommended. They are supplied

TABLE 1.

DUTIES OF IRONAC PUMPS

Pipe Connections	1½			2			3			4		
Gals. per min. 1·5 H_2SO_4	27			44			155			220		
Stages	2	3	4	2	3	4	2	3	4	2	3	4
R.P.M.		1380			1280			980			870	
Head	22	32	42	22	32	42	22	32	42	22	32	42
HP.	·68	1·02	1·36	1	1·5	2	2·6	3·9	5·2	3·8	5·7	7·6
Efficiency	40	39	38	44	42·6	42	59	57·5	57	58	56·5	55·5
R.P.M.		1800			1700			1300			1150	
Head	40	60	80	40	60	80	40	60	80	40	60	80
H.P.	1·1	1·65	2·20	1·70	2·55	3·4	4·8	7·2	9·6	6·6	9·9	13·2
Efficiency	44·5	44·3	44·5	47·4	45·4	47·1	58·5	58·4	58·5	60·5	61	60·5
R.P.M.		2160			2040			1560			1380	
Head	56	84	112	56	84	112	56	84	112	56	84	112
H.P.	1·44	2·16	2·88	2·30	3·45	4·60	6·4	9·6	12·8	9	13·5	18
Efficiency	47·6	47·6	47·5	48·6	45·8	48·7	63	62	61	62	62·5	62·3
R.P.M.		2460			2320			1770			1580	
Head	74	110	146	74	110	146	74	110	146	74	110	146
H.P.	1·90	2·85	3·70	3·1	4·6	6·2	8·8	13·2	17·6	12	18	24
Efficiency	47·5	47	48·5	48	47·8	47·8	59	58·6	58	62	61·4	61

in sizes from 1¼ in. dia. to 6 in. dia. suction pipe, with capacities ranging from 22 gals. per min. to 330 gals. per min. For some purposes pumps with regulus metal impellers are also suitable, but the rotors have to be of substantial design.

Earthenware Centrifugal Pumps.—Great improvements have recently taken place in toughening earthenware for making pumps, impellers and casings. Notwithstanding the world-wide quest for a real acid-resisting and passive metal or alloy

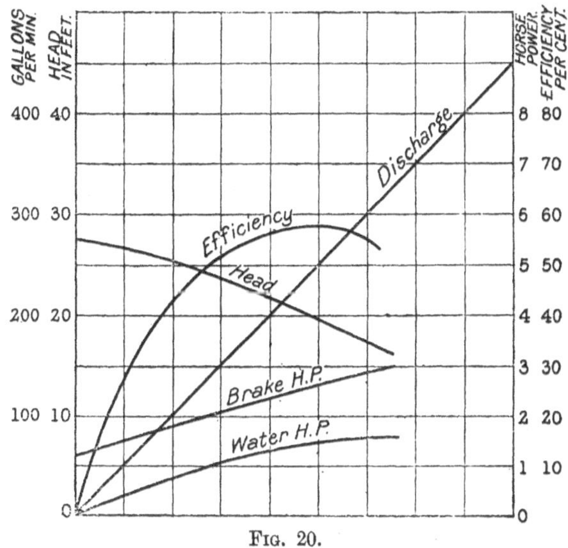

Fig. 20.

there are many and increasing classes of liquid which cannot be handled in metallic or ebonite plant. Solutions containing hydrochloric acid and soluble silver at boiling temperature, organic acids to be free from iron and other impurities, peroxide solutions, and perhaps one of the most corrosive solutions known—acid ferric chloride, demand the use of an absolutely insoluble material, capable of withstanding great changes of temperature and a certain amount of rough usage. For such purposes stoneware centrifugal pumps have been used with excellent results, and in England a pump with a ceramic lining of " Ceratherm " is marketed by Guthric & Co., Accrington.

Rubber Centrifugal Pumps.—Most chemical pump makers will supply centrifugal pumps covered and lined with ebonite and soft rubber for handling acid and gritty liquids. Among such pumps the " Stereophagus," made by The Stereophagus Pump & Engineering Co., Ltd., occupies a prominent place. The " Resiline " pump is specially designed for handling corrosive substances. In it the impeller consists of a cast-iron plate, the boss of which screws on to the spindle of the pump. On the plate is vulcanized a layer of pliant rubber from which spring curved blades unshrouded. The casing of the pump is an ordinary volute chamber which can if necessary be completely lined with rubber. This type of pump has found extended use in pumping acid mine waters containing large quantities of quartz. There is no tendency for centrifugal force to bend the vanes, because the rubber has the same specific gravity as the water, and where a fibrous piece of material enters the blades bend and permit the obstruction to pass. The diagram Fig. 20 gives the characteristics of a 4 in. Flexala pump. Compared with an ordinary centrifugal sand pump for pumping crushed gold ore, the yearly upkeep charges are from $\frac{1}{3}$ to $\frac{1}{2}$.

VI

ROTARY PUMPS

For pumping viscous solutions, liquids containing precipitates and non-gritty magmas such as the paper-maker's "stuff," the rotary pump is now usually adopted. These pumps are the half-way house between the positive ram pump and the centrifugal. Some deliver a steady non-pulsating stream, positively and without the use of valves. Most of them work

Fig. 21.—Drum Pump.

equally well in both directions. Mention will only be made of one or two as types.

The Drum pump made by Messrs. The Drum Engineering Co., Bradford, is a well known form. Fig. 21 is self explanatory. The drum and piston are geared together and a volume equal to 2π R.A. cu. ft. is swept per revolution; where R is mean radius of the projecting rib and A the area of the rib.

The Douglas pump, designed originally for pumping milk without churning, has also found application in chemical works. Essentially it consists of a rotating disc having on the periphery and equidistant three pins. These pins engage as the disc revolves in two slots in a drum, causing the latter to rotate about a fixed centre. A half moon cam-like projection on the chamber cover separates the inlet from the outlet chamber. This pump can be obtained fitted with steam jacket for the pumping of fats, waxes, chocolate and very highly viscous solutions.

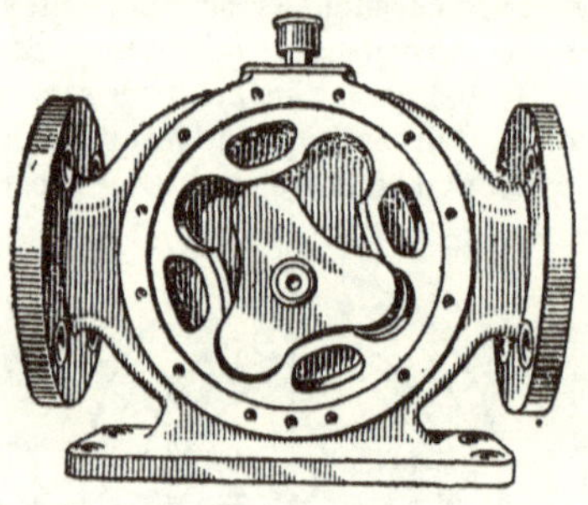

Fig. 22.—Feuerheerd Pump.

Fig. 22 shows the rotating element of another rotary pump put on the market by Messrs. Stothert & Pitt for the pumping of oils and similar liquids. The action of the pump is highly ingenious and very simple.

The pump is made in ten sizes, the smaller sizes to pump 1000 gals. of water per hour or 820 gals. of fuel oil per hour, and the larger sizes to pump 90,000 gals. per hour of water or 74,000 gals. per hour of fuel oil. The volumetric efficiencies range round 95 per cent. for heads equivalent to 200 lb. per sq. in., and mechanical efficiencies of 84 per cent. for water and 70 per cent. for heavy fuel oil. A special form is made for hydraulic jacks and control gear to pump against a head of 800 lb. per sq. in. The "Feuerheerd" pump, as it is called, is easily constructed of the new ferro silicon alloys, earthenware, porcelain and the usual restricted class of acid resistant materials. Such pumps are now obtainable having efficiencies of the values for water. Fig. 23 gives curves of volumetric (VE), mechanical (ME) and overall efficiencies (OE) also B.H.P. at different pressures and speeds. This pump is now widely adopted for the pumping of heavy oils so that it should be welcomed in many chemical industries.

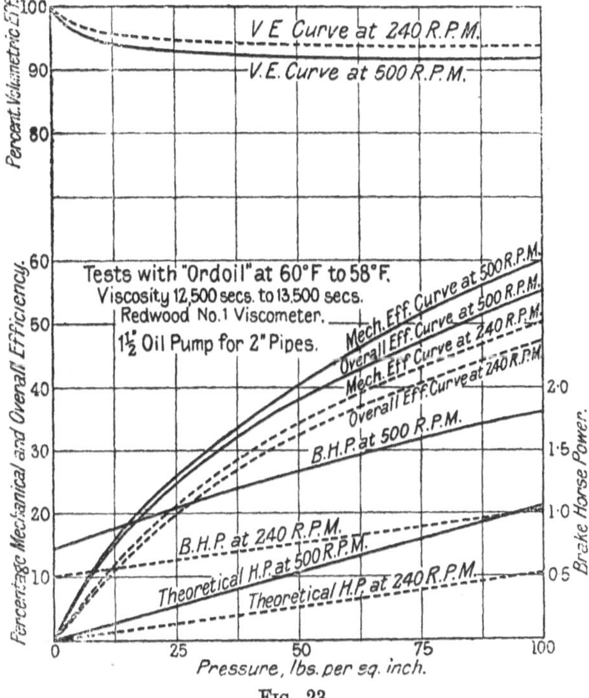

Fig. 23.

In this class is the sud pump which is now generally used to pump small quantities of soap and cutting solution on to the work on the lathe and machine tools. Most of these consist of a pair of miniature mangle gears working in an elliptical chamber. Their action is obvious from the Fig. (24) which illustrates the form made by Messrs. The Brooke

Tool Manufacturing Co., Ltd., Birmingham. Made in suitable materials it can be used for the pumping of small quantities of

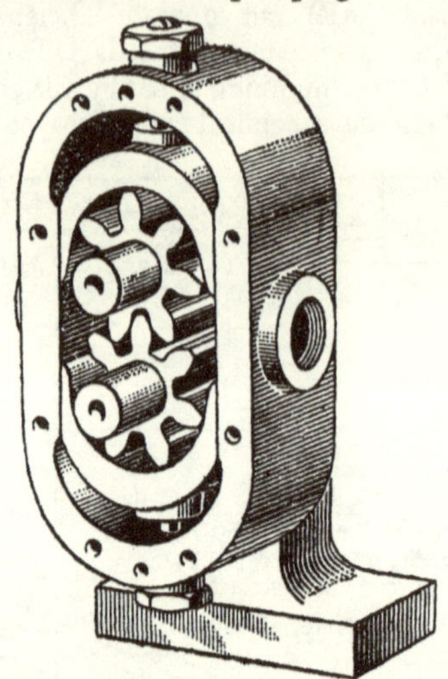

FIG. 24.—SUD PUMP.

corrosive solution without the use of storage tanks.

The Frenier Sand Pump.—The Frenier Pump is made by

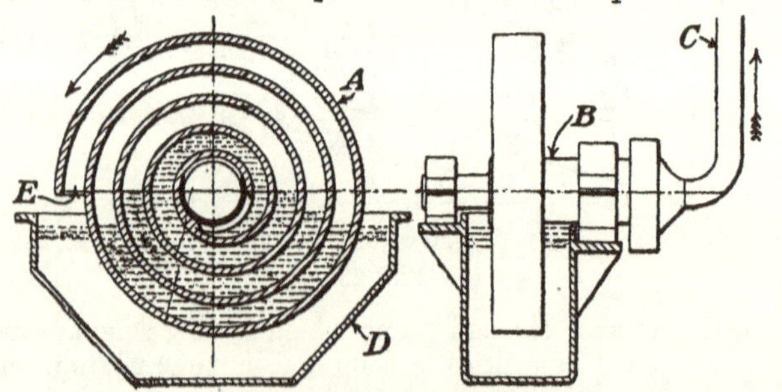

FIG. 25.—FRENIER PUMP.

Messrs. Frazer & Chalmers, of Erith, for the pumping of slimes and ore pulps. Its construction is exceedingly simple, for it consists only of a spiral chamber mounted on a trunnion in such a way that alternate slugs of air and pulp are shot out through the trunnions. The illustration (Fig. 25) best explains the construction and action of the pump. The spiral A is mounted on a trunnion B and fitted in a tank D. As the spiral revolves in the direction of the arrow the open end E traps a quantity of air on entering the liquid. Thus alternate volumes of air and liquid are wound round towards the centre, the air meanwhile being compressed. On reaching the centre the liquid is ejected through the trunnion into the pipe system C. Its characteristics are that its output is fixed because it cannot be run more than 20 r.p.m.; the height of lift is controlled by the diameter of the spiral, the efficiency demands the maintenance of a constant level in the tank. To obtain this very often the tank has to be over supplied. Table 2 gives the normal sizes and output in terms of pulp of varying specific gravity.

TABLE 2.

FRENIER PUMPING CAPACITIES

	Specific Gravity of Fluid.					
Size, Inches	1000	1100	1200	1300	1400	1500
	Tons per 24 hours of pulp.					
44 by 6 48 by 6 54 by 6	450	490	540	580	630	670
44 by 8 48 by 8 54 by 8	540	590	650	700	760	810
44 by 10 48 by 10 54 by 10	630	690	760	820	880	940

VII

GAS AND AIR PRESSURE PUMPS

UNDER this heading in our classification are placed the Humphrey pump, the ordinary acid egg, Kestner's elevator and the many similar forms of automatic egg or montejus.

The Humphrey Pump.—This highly ingenious pump has not, so far as the writer is aware, been adapted for pumping corrosive liquids as it is essentially suitable for large output, but there seems to be no reason to the contrary if the combustible gases and their products do not react with the solution pumped and if the pump is constructed of suitable materials.

An analysis of its thermodynamics appeared in *Engineering*, February 11, 1921, by Dr. John Walker.

The Humphrey pump has been taken over by Messrs. Wm. Beardmore & Co., Ltd., of Glasgow.

Acid Egg.—The egg is a large scale laboratory wash bottle. It is probably the simplest of all known acid pumps. This simplicity, however, demands personal attention, for the operation of valves is accompanied by low efficiency. Fig. 26 shows its general arrangement. The vessel A is normally about 3 ft. in diameter by about 6 ft. long and contains 40 cu. ft. of acid. The delivery pipe B extends to the bottom of the vessel and dips into a small well C. Air is admitted by the cock D, and the acid to be lifted by the cock F from the tank E. To operate the vessel A is filled with acid, the displaced air issuing from the cock G. When full, air is admitted after first closing cocks F and G, when the acid is forced up pipe B. After the operation it is seen that the vessel A is full of compressed air at approximately the *maximum* pressure, the energy of which is entirely lost by expanding up the delivery pipe B. Thus the working of the acid egg is exceedingly crude. By perforating that portion of the discharge pipe inside the egg it is possible to raise the acid by using air expansively, but the degree of efficiency depends on the intelligence of the attendant. The study of the egg has been left too much in the hands of the air-compressing engineer. It is true, of course, that the ultimate cost of raising acid by

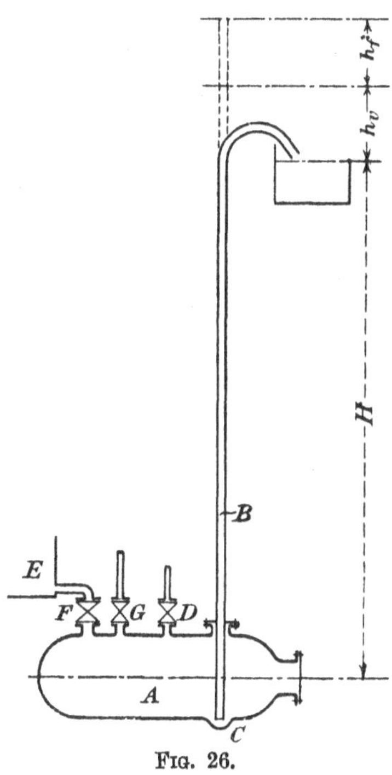

FIG. 26.

this means depends on the efficiency of the air compressor; but the egg should be studied as an appliance independent of the compressor.

Let $H =$ the actual head to which the acid is raised.
and $h_f =$ friction head in feet.
$h_v =$ velocity head in feet.

d = dia. of rising main in feet.
Q = volume of acid raised cu. ft. per sec.
v = velocity of discharge feet per sec.
m = hydraulic mean depth of rising main $= \frac{d}{4}$
i = hydraulic slope $\frac{h_f}{H}$
ν = kinematic viscosity (for sulphuric acid 1·84 sp. gr. at 20° C. ν = ·000113 ft. sec. units).
A = area of rising main sq. ft.

Maximum pressure of air required :—

$H + h_f + h_v$ in feet of the liquid being pumped.

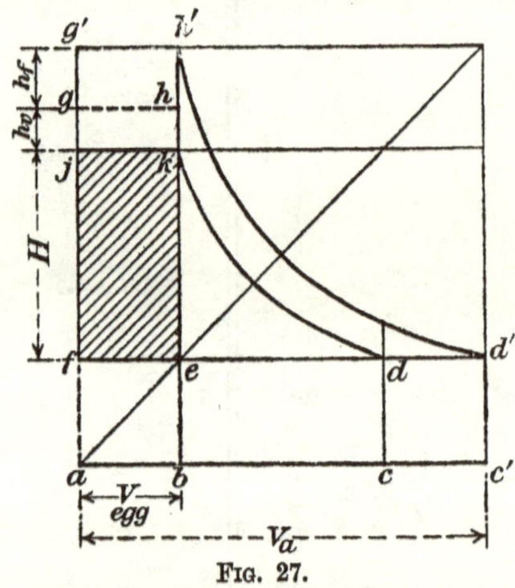

Fig. 27.

Velocity head $h_v = \frac{v^2}{2_g}$ where $v = \frac{Q}{A}$

Friction Head h_f is calculated from the Stanton curve (*Flow of Chemical Liquids*, Swindin, in this series) :—

i is obtained from

$$i = \varphi \frac{v^2 vd}{m\nu}$$

and h_f can therefore be calculated.

These values are given diagrammatically in Fig. 27.

ab represents volume of egg.

af = atmospheric pressure in terms of head of liquid raised (for sulphuric acid 1·84 sp. gr. $af = 18{\cdot}4$ ft.).

$fg' = \mathrm{H} + h_f + h_v$.

At the end of the blow the egg contains a volume of air ab at pressure ag' which expands to the atmosphere along the line $h'\ d'$. Actual work done on the acid is represented by area $f\,j$ K e; work done on the air area $f\,g'\ h'\ d'$; work done to overcome frictional resistance and cause flow is area $g'\ h'\,j\,k$.

The efficiency of the egg therefore is given by the ratio

$$\frac{\text{area } f\,j\,k\,e}{\text{area } f\,g'\,h'\,d'}$$

In practice this does not exceed 40 per cent., and assuming a compressor efficiency of 60 per cent. the overall efficiency is then 24 per cent. If the attendant does not keep his ear well glued to the rising main during the discharge period loss of air can be anything. The writer generally regards the practical efficiency as anything between 10 per cent. and 20 per cent.

The following table has been worked out for varying velocities of discharge. From the foregoing it can be seen that efficiency depends on low value of v.

The following diagram, Fig. 28, gives graphically the chief factors of the acid egg action. It is based on an egg of 1000 gallons capacity, and a lift H of 40 ft. Referring to quadrant (1) the egg and its rising main are shown in outline, the head H and the length of the egg forming the $p\,v$ rectangle of the expansion hyperbola. This rectangle A B C D indicates the work actually done on the acid. To cause flow further work is required, and which is indicated by the lengths h_v, h_f added to H. The intercept of the isothermal curves, as at G, give the volume of free air required for raising 1000 gallons to a height H, at the velocities stated. Quadrant (2) contains a curve WD connecting the air pressure in the egg with the velocity of pumping. The scale of velocities and gallons per minute is given on the base line, and the extreme ordinate contains the scale of pressure. Quadrant 3 gives the corresponding H.P. for the velocities; AK the actual H.P. absorbed in actual pumping; AL the H.P. to be given to the air, the cross sectioned area indicates the H.P. wasted on expansion, MA and NA show the H.P. for compressors of 80 per cent. + 60 per cent. respectively. The curved line J converts velocity or gallons per minute into time required to pump 1000 gallons shown on scale to the left. Quadrant A contains the efficiency values. Efficiencies of H.P. on line KA Quadrant 3, are given by 100 per cent. efficiency line (vertical), H.P. LA on line V, H.P. MA on line U, and H.P. NA on line RP. The cross curves P, TUV, R and S indicate efficiencies for velocities 3, 6, 9 and 12 ft. per second respectively. The sectioned area Quadrant A shows at a glance the scope offered for expansive working. The diagram as a whole refers only to the case given, 1000 gallons raised 40 ft. For other heads a similar diagram must be constructed. The values from which the chart was drawn are given in Table 3.

TABLE 3.

ACID EGG

Data of Flow Values for 1000 *gals. of* 1·84 *Sulphuric Acid through a* 2 *in. lead pipe, raised* 40 *ft. from an acid egg.*

Velocity	0	3	6	9	12
Actual head lifted	40	40	40	40	40
Head velocity *hv*	0	·14	·57	1·27	2·25
Head friction *hf*	0	1·4	4·32	8·52	13·0
Head total H + *hv* + *hf*	40	41·54	44·89	49·79	55·28
Efficiency on air only	55	—	47·4	42·0	37·0
Efficiency 80 per cent.	44	—	37·92	33·6	29·6
Efficiency 60 per cent.	·33	—	28·4	25·2	22·2
H.P. per 1,000 gals. — Theoretical — Acid	—	·55	1·1	1·66	2·2
H.P. per 1,000 gals. — Theoretical — Air	—	—	2·32	3·93	5·94
H.P. per 1,000 gals. — 80 per cent. compression	—	—	2·9	4·9	7·44
H.P. per 1,000 gals. — 60 per cent. compression	—	—	3·86	6·54	9·9
Pressure lb. per sq. in.	32	34	35·9	39·7	44·2

Efficiency in a chemical works has not hitherto been of great importance for the pumping of small quantities of liquids, so that it is not surprising that inventors have concentrated on the elimination of the personal element. Mention may be made of the well known Kestner elevator, which in these days needs no description. Fig. 29 shows an automatic

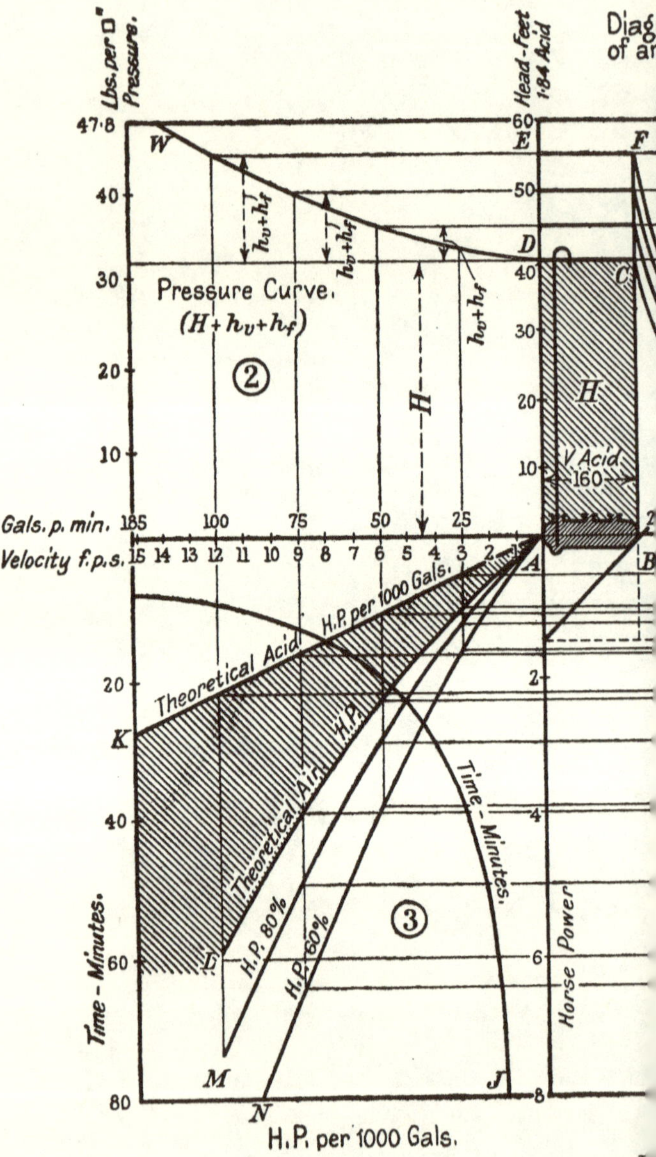
Diagr
of an
Lbs. per □"
Pressure.
Head-Feet
1·84 Acid
Pressure Curve.
$(H+h_v+h_f)$
②
h_v+h_f
H
V Acid
160
Gals. p. min.
Velocity f.p.s.
H.P. per 1000 Gals.
Theoretical Acid
Theoretical Air
H.P.
H.P. 80%
H.P. 60%
Time - Minutes.
③
Horse Power
Time - Minutes.
H.P. per 1000 Gals.

howing Efficiency, H.P., and Air Volumes
. Egg blowing 1000 Gallons of 1·84 Sp. Gr.
Sulphuric Acid.

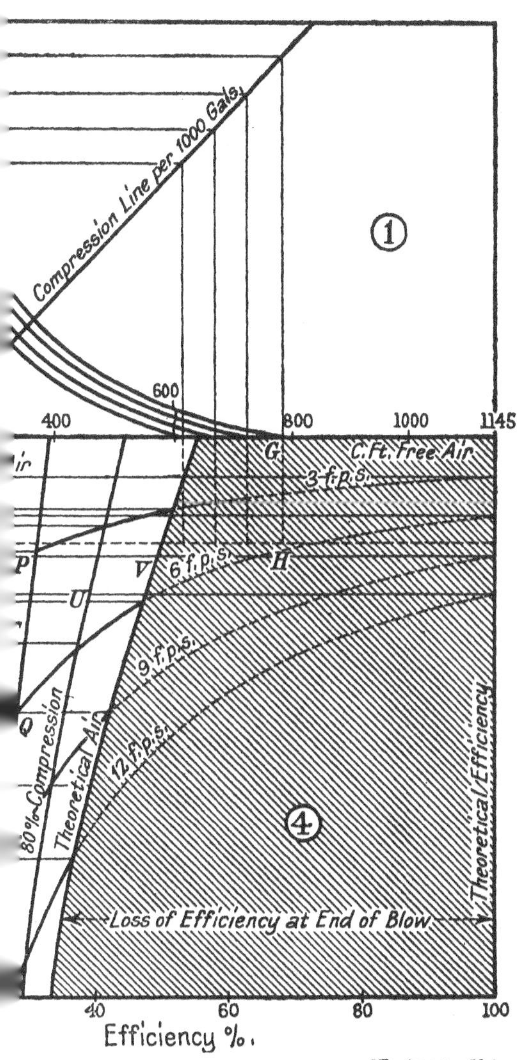

[To face p. 52.]

egg in Narki iron, by Barnes, and supplied by Messrs. Varley, of St. Helens, in which the usual float is replaced by an internal valve. The air enters by the side branch of the valve

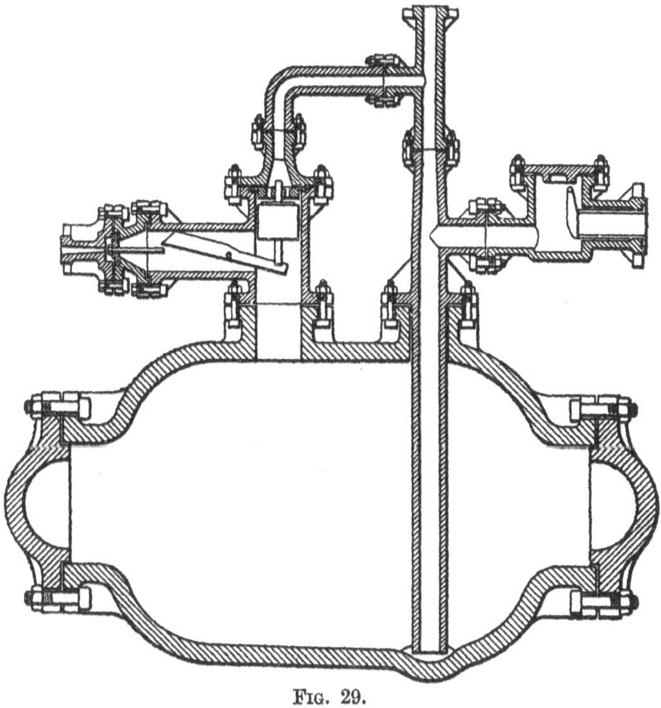

Fig. 29.

chamber, as soon as the egg is filled with acid, and the float valve is on its seat. The acid is forced up the rising main till the pressure falls, and closes the air inlet valve. A large variety of similar devices may be obtained.

VIII

MISCELLANEOUS DIRECT-ACTING LIFTING APPLIANCES

The Screw Pump or Archimedean Spiral.—This appliance is met with in two forms, one consisting of a coarse pitched screw rotating in a cylinder fixed at any angle, the other of

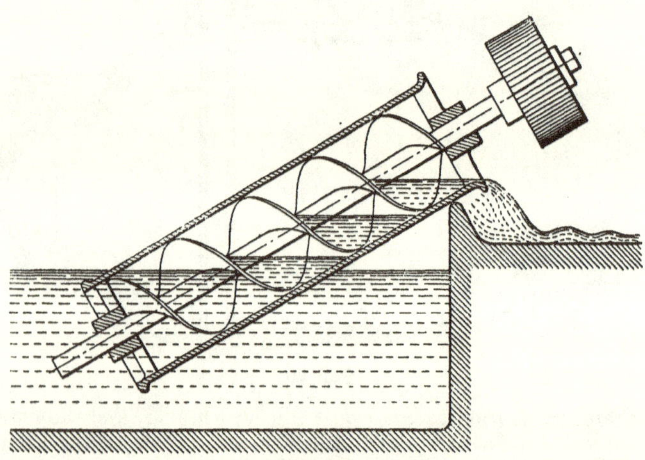

Fig. 30.

a pipe coiled helically and fitted round a shaft capable of rotation. In up-to-date hydraulic schemes it is rarely seen, but the chemical engineer can still use this simple and practically gearless pump. It is employed on some electrolytic cells.

The design of these pumps does not call for much effort. For the rotating screw:

If d denotes the dia. of the cylinder.
p ,, ,, pitch of the screw.
v ,, ,, volume of screw length of p.
n ,, revs. per min.
ϱ ,, density of material pumped.
K ,, co-efficient for slippage and friction losses.
Q ,, mass of liquid raised.
h ,, head

then:

$$Q = \frac{\pi d}{4} \times p \times n \times \varrho$$

and work done:

$$= Q \times K \times h.$$

For practical purposes K is equal to 0·8, an efficiency of 80 per cent. for water. When pumping mercury or molten lead and similar alloys it must be remembered that loss due to leakage is much higher. Though densities and viscosities are greater than those of water the leakage is a function also of surface tension. Those who have experience of keeping a concrete tank tight against mercury will appreciate this point. Anyhow the point of efficiency in the chemical works need not be laboured, for the quantities of liquids to be dealt with do not compare with a town's water or sewage. Again, where possible, adopt the rotating screw form in preference to the helical pipe. A screw has no inaccessible parts and can be withdrawn for cleaning; a pipe is not so easily cleaned, especially when twisted. This appliance is often used as a liquid circulating device and is frequently seen in the various forms of electrolytic cell. Figs. 31 and 32 give two examples of the Archimedean screw used as a circulator, one vertical, the other horizontal.

The Scoop Wheel.—Where large volumes of liquors require to be circulated in shallow tanks the scoop wheel has advantages over pumps of any other kind. It has been employed in circulating molten alloys, electrolytes in electrolytic cells, and also for strong sulphuric acid. The wheel consists of a Sagebien water wheel reversed and is a favourite water raiser

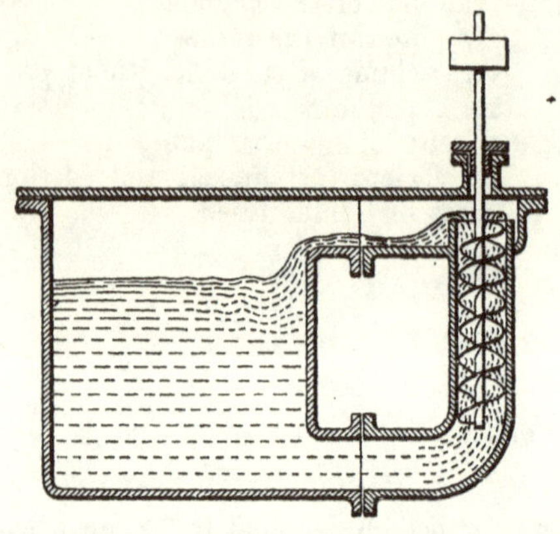

Fig. 31.—Vertical Circulator (Archimedean Screw).

of the Dutch, who use it to keep in motion the water of their sluggish canals.

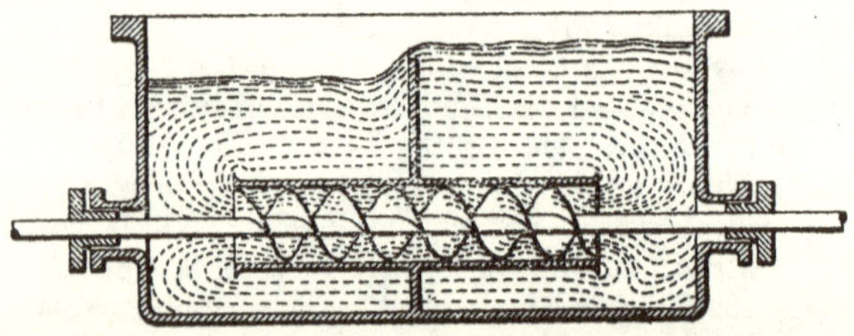

Fig. 32.—Horizontal Circulator (Archimedean Screw).

Fig. 33 shows the general construction of the wheel. In designing such an appliance the reader must consult a competent work on hydraulics and modify details for the liquid to be pumped. It will of course be made of suitable material, hard lead for cold sulphuric, ebonite-covered steel for hydrochloric acid, and so on. The angles of the floats are usually made 20° to 40° with the radial line, and the diameter of the wheel lies between 9 and 10 where H is lift in feet. The peripheral speed will depend on the density and viscosity of the liquid. For water this speed is 8 ft. per sec. It is

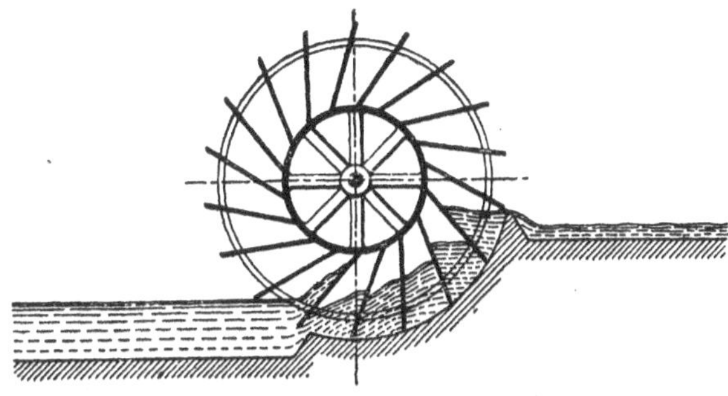

Fig. 33.

here suggested that for other liquids this speed of 8 ft. per sec. be multiplied by the factor:

$$\frac{\nu \text{ for water}}{\nu \text{ for liquid to be pumped}}$$

where ν is kinematic viscosity $\frac{\eta}{\varrho}$, where η is absolute viscosity c.g.s. units and ϱ density c.g.s. units. Owing to its moderate speed hydraulic losses are very low and in consequence the efficiency is high, about 80 per cent.

IX

THE JET PUMP AND BLOWER

It is not necessary in this small work to go deeply into the action of jets and nozzles, but a right understanding of the jet is important, for on it depends the design of steam turbines, steam injectors, both for gases and liquids, high flow pressure gas and oil burners, and nowadays the high vacuum condenser air pump. In the chemical works the jet is met with as boiler feed pumps, forced draught appliances for burning difficult fuels, burners for the combustion of tar, fuel oil and gas, and as a rough and ready draw-off pump.

The basic principles of the jet are usually treated as problems of colliding bodies in space. If a body of mass M, with a velocity V, strikes another body at rest of mass M_2, the common velocity is

$$V = \frac{M_1V_1}{M_1M_2}$$

and the ratio

$$\frac{\text{lost energy}}{\text{remaining energy}} = \frac{M_2}{M_1}$$

From this it is seen that the greater the stationary body, the greater is the loss. Thus in a simple type of jet the efficiency is very low. Now jets may be roughly divided into two classes. 1. Those in which the motive fluid is either a liquid or a non-condensible gas. 2. Those in which the motive fluid condenses like steam. In class 1 the jet velocity is that due to the pressure above the atmosphere. In the latter case sudden condensation so reduces the pressure that jet velocities of about 1,600 feet per second are obtainable. Thus it is that exhaust steam can pump water into the boiler, and also that a vapour of small mass can move a liquid of much greater mass. It is not practical to move liquids by air issuing from a jet, except in sprays and the like where mechanical efficiency is not looked for.

The steam ejector or blow jack, as illustrated in Fig. 34, of the chemical works is the crudest appliance imaginable.

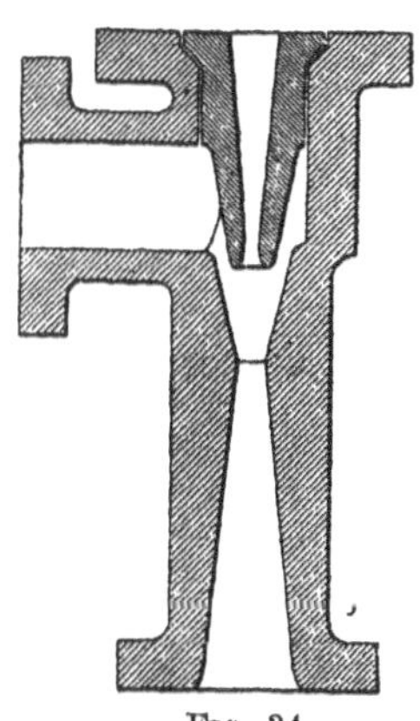

Fig. 34.

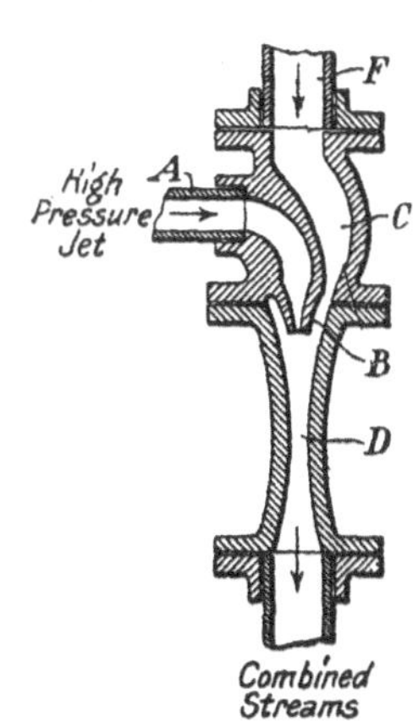

Fig. 35.

It should only be used for intermittent work, as emptying pumps and vats at long intervals, where the capital cost of an efficient pump would be prohibitive. Its efficiency is improved if fixed so that the work is all done on the suction side, and not on the delivery, and also if the jets are made to draw the liquid through telescopic openings like those by Korting, Ltd.

The use of jets for pumping gases, as in fuel burners, is not discussed here.

A form of jet pump (Fig. 35), known as the Thompson pump, is used occasionally where a small volume of one liquid at a high pressure, entering the jet, causes the motion of a greater volume of liquid at a lower velocity. For chemical works, fire service, or for dewatering foundations, it may be of service.

X

AIR LIFT PUMPS

Pohle of America (1886) is generally credited with the first true conception of the working of the air lift, but this method of raising liquids was known earlier. Carl Löscher probably invented it in the year 1797. After a premature burial it seems at last to have arisen to witness a partial eclipse of other pumping devices. From being the last resort of a baffled engineer to make a true acid-resisting pump, and one that can pump sandy water from a deep small bore hole without hurting itself, it is at last being studied with that thoroughness which it deserves, and attempts are now made to get efficiencies which compare favourably with other pumps.

The air lift, however it may be constructed, is but a U tube having one limb longer than the other. The short limb contains a heavy liquid, the long limb a lighter liquid. The heavy liquid is the liquid to be pumped in a homogeneous condition, the lighter liquid is the same amount of substance aerated by the addition of air or gas. Fig. 36 shows an elemental air lift in which AB is the short limb receiving the liquid to be pumped, CD the long limb or rising main containing the aerated liquid. The pipe EC delivers the air or gas to the footpiece at C of this long limb.

Though the lift is so simple in construction its action is "wropt in mystery." There is at present no adequate theory developed upon which to base design. Its output is controlled entirely by undefined resistances such as slip of bubbles through the liquid, fluid friction of a mixture of liquid and gas, and eddies.

In order to calculate the factors for equilibrium for a given lift it is usual to assume that the air expands isothermally, i.e. to say no one has yet published any data concerning the action of the lift when pumping hot liquors or even molten metals. In the latter case expansion of the air is a complex affair, for it is neither isothermal nor adiabatic, for the air is receiving energy from the heat of the liquid being pumped. The action of the lift is complex enough without thinking at the moment of the effect of added energy, so we will consider the lift of a liquid at the temperature of the atmosphere.

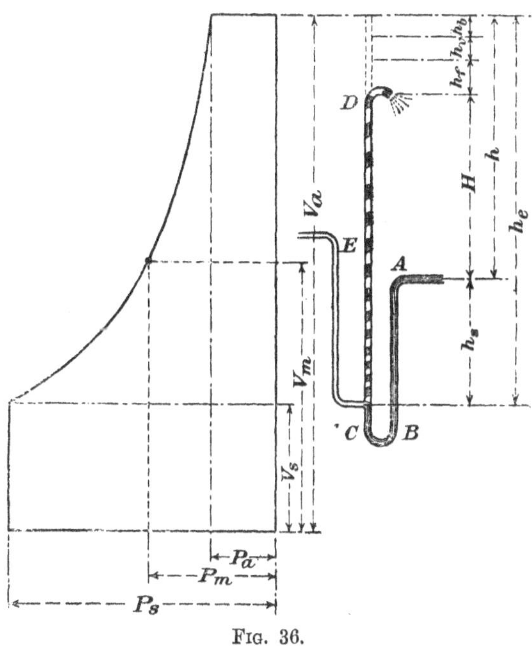

FIG. 36.

Let H = the actual lift in feet.

h_s = the submergence in feet.

h = the theoretical lift in feet (i.e. the height to which the liquid will rise but not flow out of the rising main).

V_w = the volume of water to be pumped per minute in cu. ft.

V_a = the volume of air required per minute at atmospheric pressure.

V_m = the mean volume of air expanding in the rising main.

V_s = the volume of air at absolute pressure P_s.

P_a = the absolute pressure lb. per sq. in. of the atmosphere.

P_s = the absolute pressure lb. per sq. in. of the air under submergence head h_s.

r = ratio of expansion $\frac{V_a}{V_s}$ or $\frac{P_s}{P_a}$.

In compressing the air isothermally to the pressure corresponding to the submergence h_s the work done per minute

$$= 144\ P_a\ V_a \log_e \frac{P_s}{P_a} \text{ ft. lb.}$$

The work done on the water per minute $= 62{\cdot}4\ V_w\ H$ ft. lb. To find the volume of air per minute when compressed under submergence head h_s:

Assuming the law $PV = K$

Then: $$P_a V_a = P_s V_s$$

$$V_s = \frac{P_a V_a}{P_s} = V_a \times \frac{1}{r}.$$

Ratio of expansion $= r = \frac{V_a}{V_s} = \frac{P_s}{P_a}.$

The volume of air $V_a \times \frac{1}{r}$ (pe r minute) at pressure P expands to volume V_a at pressure P_a.

The mean volume during this expansion

$$= V_m = V_a \times \frac{\log_e r}{r-1}.$$

The specific gravity of the aerated column compared with that of the liquid $= \frac{1}{1 + V_m}$, and this is in equilibrium with h_s.

$$\therefore h_s = (H + h_f + \text{head producing flow}) \frac{1}{1 + V_m}.$$

$$\therefore h_s = (h_s + h)\frac{1}{1 + V_m}.$$

$$\therefore h = h_s (V_m).$$

So far this is ordinary textbook information, and all we know is the height to which the liquid will rise but not flow out of the rising main for a given volume of air. No more energy can be supplied to the appliance, for this is governed by the submergence. Clearly then to cause flow we shall have to content ourselves with raising the liquid to some other height H less than h. Then $h - H$ is the head producing motion and overcoming all the resistance to flow. These resistances are :

Loss at entry at the foot piece.

Friction loss of liquid.

Loss at discharge of liquid at top of rising main.

Loss through slippage, that is the rise of the bubbles of air through the liquid as the aerated mass rises.

Analyses of many carefully carried out tests show that the friction in the rising mass of an air lift is about six times that for the flow of water alone at the same velocity.

Friction.—The following method for calculating friction is here suggested, though it is not possible to give values in the absence of experimental work. Stanton has shown that the relation

$$\frac{mig}{v^2} = \varphi\frac{vd}{\nu}, \text{ where } m = \text{hydraulic mean depth } \frac{d}{4}$$

$$i = \text{hydraulic slope } \frac{h}{l}$$

$$v = \text{velocity}$$

$$\nu = \text{kinematic viscosity } \frac{\eta}{\varrho}$$

$$\eta = \text{absolute viscosity}$$

$$\varrho = \text{density}$$

holds for all fluids. See Swindin, this series, *Flow of Liquids*. Now if some industrious worker will give us the figure for η for a series of air and liquid mixtures, the value of friction could be obtained with fair accuracy. It is presumed that the value for critical velocity would be much modified, for it cannot be conceived that the lowest velocity of an air and gas mixture would flow in streamline fashion.

Until such experiments are made the following modification of Stanton's formula is suggested:

Having assumed and obtained the values of

v_m = mean velocity feet per sec. of the mixture

V_m = mean volume of air in cu. ft. per cu. ft. of liquid

ϱ_m = mean density of mixture from

$$\varrho_m = \frac{\text{specific gravity of liquid}}{1 + V_m}.$$

η = absolute viscosity of liquid under conditions of flow

ϱ = density of liquid

first obtain value of ν, taking η for the pure liquid,

$$\nu = \frac{\eta}{\varrho_m}$$

with this find value of Reynolds' criterion:

$$\frac{vd}{\nu}$$

referring to chart in *Flow of Liquids*, this series, obtain corresponding value of

$$\frac{mi}{v^2} = \varphi \frac{vd}{\nu}$$

from which value of i is at once derived.

The correct value of h_f cannot be calculated directly, because the length of rising main is not yet known.

$$h_f = i\,(H + h_s).$$

The value of H, the actual lift for any set of conditions, is obtained from the equation given later on page 66.

Kinetic Head h_v.

$$h_v = \frac{v^2}{2g}, \text{ where } v = \text{mean velocity of the mixture.}$$

Loss due to Slippage.—A bubble of air introduced in a body of liquid at the bottom of the rising main will rise or slip through, its energy being dissipated in overcoming frictional resistance of the liquid. It would now be possible to calculate the velocity of movement of the air bubble with respect to the liquid, and this has been done by Daw in his work on "Compressed Air Power." The results of such theoretical calculations however are not reliable, as the size of bubble is not exactly known. A more practical method is to ascertain this essential figure by experiment.

A paper read before the British Association at Edinburgh, September, 1921, by Dr. John S. Owen, gives some results of careful experiments carried out to determine the size of bubbles issuing from pipes and orifices and their velocity when rising through a column of water in a glass tube. Owen found that the size of bubble was not a simple function of the diameter of the orifice. Very small bubbles were influenced largely by surface tension and rate of flow and that there was a minimum diameter which it was difficult to get below. The following Table (4) is extracted from Dr. Owen's paper.

TABLE 4.

Nozzle Dia.	Bubble Dia. Approx.	Velocity of Rise.
in.	in.	f.s.
0·012	0·05	0·53
0·012	0·10	0·57
0·136	0·125	0·67
0·136	0·15	0·70
0·212	0·25	0·80
0·375	0·37	0·80
0·375	0·38	0·81

It is obvious therefore that for a low value for slippage loss the bubbles should be kept small and the velocity of the rising mixture high. This latter condition gives a high value for friction, so that careful design means compromising slippage and friction losses.

The following method of calculating the value of slippage loss for water in terms of head is given tentatively pending a thorough investigation of the problem. Taking the reasonable average figure of 0·75 foot per second as the velocity of a rising bubble in water from a $\frac{1}{4}$ in. orifice: if V_m is the mean velocity of the mixture in the rising main, then allowing for slip, the actual mean velocity is $(V_m - \cdot 75)$ ft. per second. The time of flow in the rising main is $\frac{H + h_s}{V_m}$ seconds and therefore the total slip is $\frac{H + h_s}{V_m} \times \frac{3}{4}$ ft.

To apply this rule to other liquids preliminary experiments should be made to determine the rise of small bubbles in a long glass tube approximately of the diameter of the proposed lift. For the usual run of chemical substances this tube is about 1 in. to 2 in. In general terms then if v_b is the velocity of the rising bubble the loss is :

$$h_b = \left(\frac{H + h_s}{v_m}\right) v_b$$

To obtain H the actual lift in order to evaluate the expressions :

$$\text{Friction head } h_f = i\,(H + h_s)$$

$$\text{Slippage head } h_b = \left(\frac{H + h_s}{v_m}\right) v_b$$

the following equation is given :

If h_e be the total length of the rising main then $h_e = (h + h_s)$ and $h_b + h_s + h_f + h_v + H = h_e$.

Substituting the above values for h_b, h_f and h_v, we get

$$\left(\frac{H + h_s}{v_m}\right) v_b + h_s + i\,(H + h_s) + h_v + H = h_e.$$

An example of this is worked out on p. 72.

From what has already been stated it follows that the air lift considered as a thing in itself is as efficient as any ordinary pump. Properly installed it converts 80 per cent. of the energy supplied to it into useful work. Having claimed this and so given it an air of respectability we can now study some of its special characteristics which ought to appeal to a chemical engineer.

In the first place there are two types of air lift : one is the pulsating type, described in Pohlé's original patent specification, in which the aerated column consists of plugs or pistons of liquid separated by bubbles, or really cylinders of gas ; the other is the emulseur type, in which the aerated column consists of a homogeneous mixture of liquid and very small bubbles of gas. The engineer's form of the lift is the emulseur or non-pulsating type, while the small lift used in the chemical works, e.g. on nitric acid towers, is the pulsating piston type. This choice of types arises from the

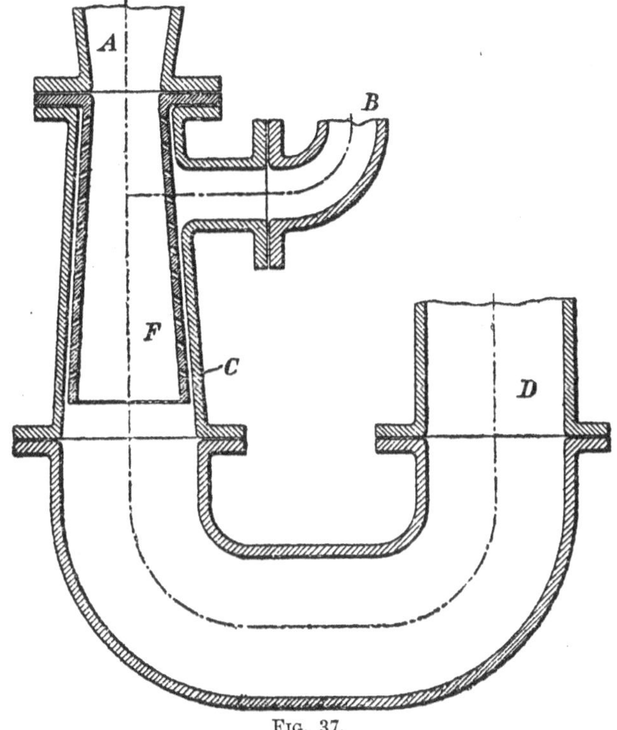

FIG. 37.

different conditions under which the lift operates. The engineers' lift is used chiefly for pumping large quantities of water and oil from deep bore holes and is constructed of large diameter pipes generally 4 in. to 12 in. diameter, too large for the air to form bubbles of the diameter of the pipes. The chemical lift is installed for lifting comparatively small

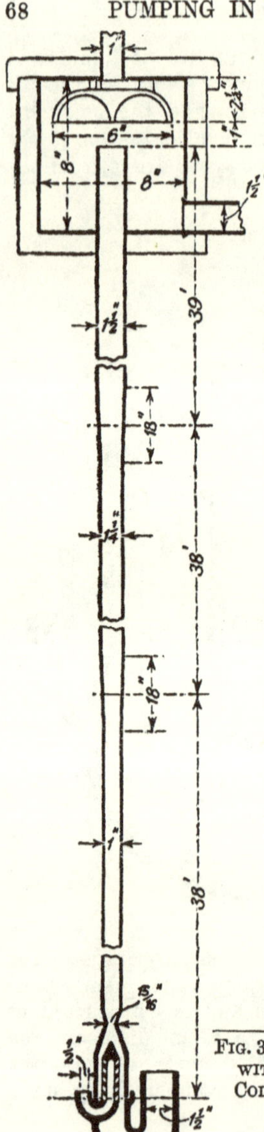

quantities of liquids in pipes generally less than 2 in. in diameter in which the air which comes from an open-ended pipe readily forms bubbles, which at once stretch across the tube separating the liquid into a succession of pistons. It has been laid down that for high efficiency the air stream should be finely divided and the liquid thoroughly aerated by small bubbles because loss arising from slippage diminishes with the smallness of the bubbles. It is evident, therefore, that the emulseur type of lift is more efficient than the pulsating Pohlé type. It may seem paradoxical to state that the very simplicity of the lift has led to its own undoing. A well, a simple pipe for rising main and another pipe as simple but smaller as air pipe may constitute a lift, but it will not be efficient. If the pipe is large enough it may work as an emulseur, but small pipes so arranged will certainly produce the Pohlé effect. It is possible to operate the smallest lift in the emulseur fashion, and to do this it is essential to divide the air so that the bubbles will not fill the rising main. Now the size of a bubble in a small lift is a function of surface tension and not simply that of the orifice through which the air passes. Dr. Owen's experiments have shown that for

Fig. 38.—Standard Lead Acid Pump with Umbrella and Lead-lined Collector Head by Sullivan.

below a certain diameter of orifice the bubble formed was always larger than the opening. If now a half-formed bubble could be swept from the orifice by the flow of the liquid a smaller bubble would be formed. A realization of this fact enables a rational footpiece to be designed. Fig. 37 (p. 67) shows the principles upon which the best type of footpiece is designed. Referring to this figure, D is the submergence pipe, C footpiece containing F a perforated Venturi throat. Air enters at B

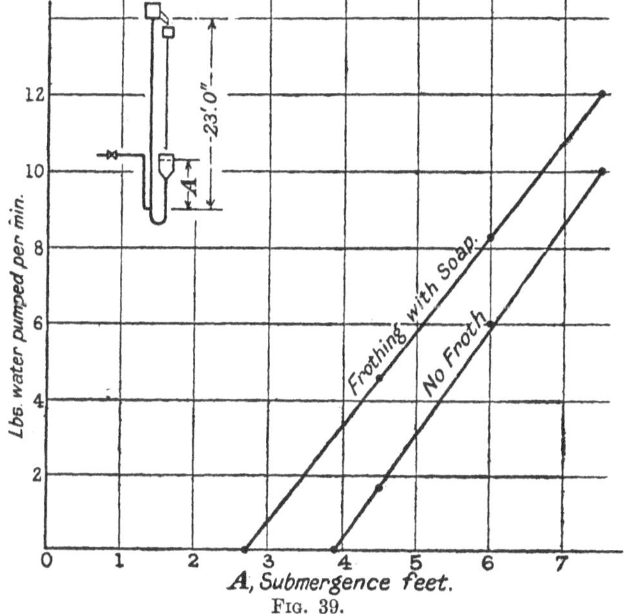

Fig. 39.

and passes in a finely divided form through the perforations of the throat into the rising main A. This design enables the air to enter a rapidly moving stream of liquid at the point of diminished pressure. Two objects are thus attained, one the brushing off the air bubbles before they attain the maximum size and which also prevents coalescing, the other an increase of head forcing the air through the fine openings in the throat on account of the Venturi effect. Fig. 38 (p. 68) shows forms of footpiece and air lift designed by Messrs. Sullivan Machinery Co. of Chicago for use in chemical works.

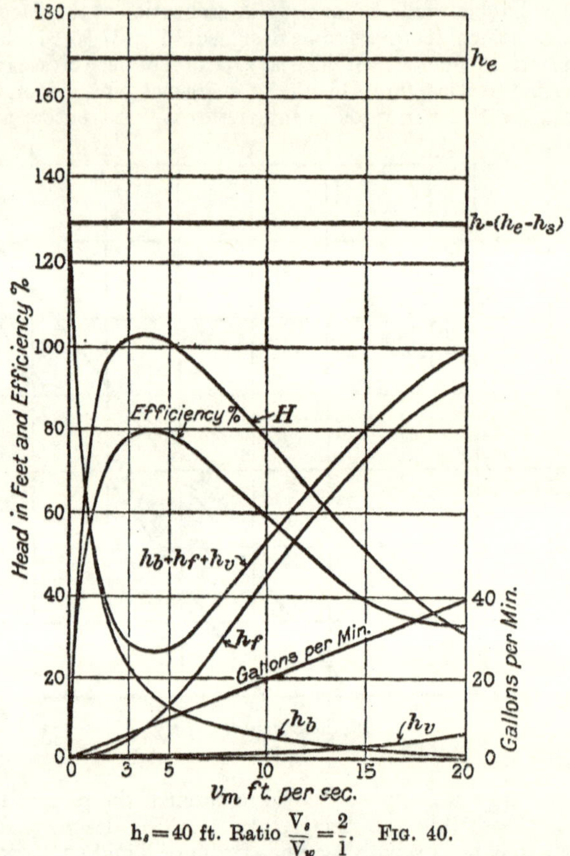

$h_s = 40$ ft. Ratio $\frac{V_s}{V_w} = \frac{2}{1}$. Fig. 40.

The Venturi construction appears immediately above the perforated air discharge pipe. It is evident that the principles above described are partly embodied in this design.

Heavy viscous liquids are easier to raise by air lift than

water. An increase of density increases in direct proportion the submergence pressure. Viscosity determines the degree of slip of the bubble, the greater the viscosity the less the slip. As resistance to flow is a function of $\frac{vd}{\nu}$, where ν is the ratio $\frac{\text{viscosity}}{\text{density}}$, it follows therefore that the aeration of the ascending column by reducing the density increases the value of ν and decreases $\frac{vd}{\nu}$.

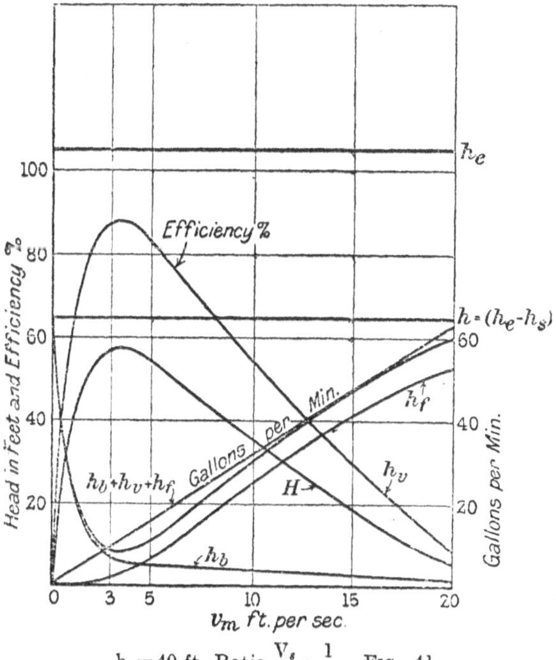

$h_s = 40$ ft. Ratio $\frac{V_s}{V_w} = \frac{1}{1}$. FIG. 41.

The effect of a change of surface tension is seen in the diagram, Fig. 39, on p. 69, which shows the quantities of water pumped for given submergence with and without the addition of soap, just sufficient to produce a slight lather. This increase in quantity, due to the lowering of surface tension, is more apparent at high ratios of lift/submergence.

Diagrams, Figs. 40, 41, 42 and 43, pp. 70–73, have been worked out for the pumping of sulphuric acid of 120° Tw.

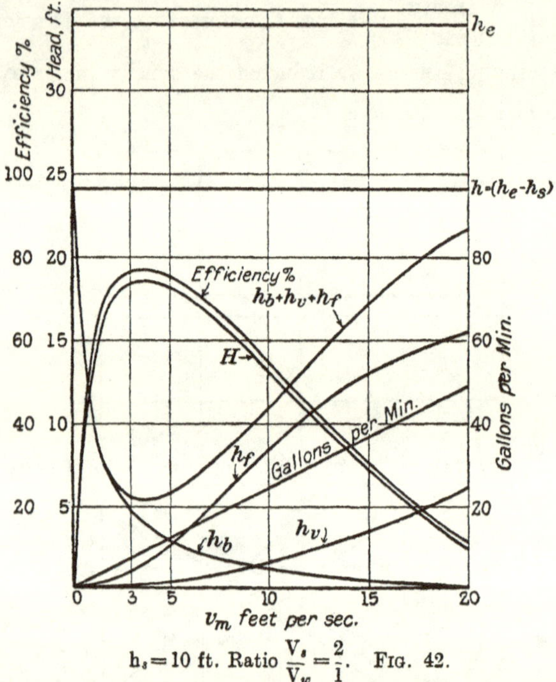

$h_s = 10$ ft. Ratio $\frac{V_s}{V_w} = \frac{2}{1}$. Fig. 42.

(1·6 sp. gr.) from the formulæ already given. By way of explanation the calculations for one particular set of conditions will be given in detail. Data :

Liquid = Sulphuric acid 1·6 sp. gr.
v_m = 3 ft. per sec.
Pipe = Lead 2 in. inside diameter
Air = 1 cu. ft. of compressed air per cu. ft. of liquid raised = V_s

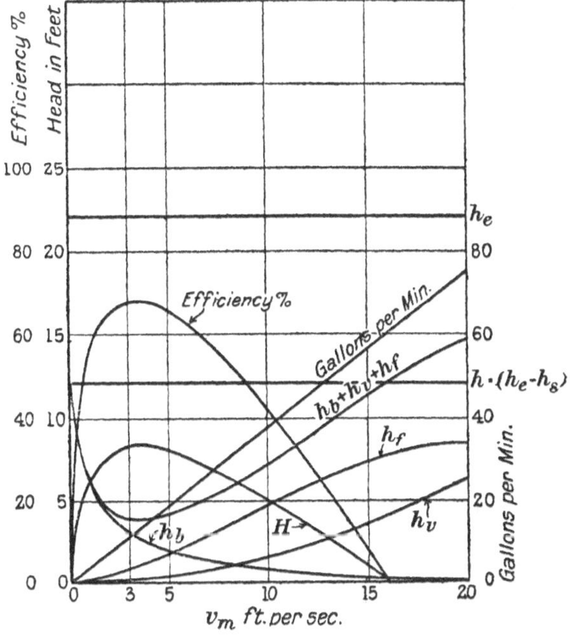

$h_s = 10$ ft. Ratio $\frac{V_s}{V_w} = \frac{1}{1}$. FIG. 43.

Submergence $h_s = 30$ ft.

Calculations :

A column of acid equal to the pressure of the atmosphere is $\frac{34}{1 \cdot 6} = 21 \cdot 2$ ft.

and r then is $\frac{21 \cdot 2 + 30}{21 \cdot 2} = 2 \cdot 42$

$$\log_e r = 0 \cdot 884$$

$$V_m = \frac{rVs}{r - 1} \log_e r = \frac{2 \cdot 42 \times 1}{1 \cdot 42} \times 0 \cdot 884 = 1 \cdot 505.$$

Mean density ϱ_m is $\frac{1 \cdot 6}{1 + V_m} = \frac{1 \cdot 6}{1 + 1 \cdot 505} = 0 \cdot 639$

Theoretical lift h is given by the equation

$$h + h_s = h_s (1 + V_m)$$

whence $$h = 45{\cdot}15 \text{ ft.}$$

Velocity head in terms of rising main: $h_v = \frac{v^2}{2_g}$

$$= \frac{9}{64} = 0{\cdot}14 \text{ ft.}$$

Friction head in terms of rising main h_f:

$h'_f = i\,(\mathrm{H} + h_s)$.

To obtain i first obtain value of $\frac{vd}{\nu}$

$$\nu = \frac{\eta}{\varrho} = \frac{0{\cdot}10}{0{\cdot}639} = 0{\cdot}156 \text{ in c.g.s. units}$$

$$= \frac{0{\cdot}156}{929} = {\cdot}000168 \text{ in ft. sec. units,}$$

for which $\frac{vd}{\nu} = \frac{3}{1} \times \frac{1}{6} \times \frac{1}{{\cdot}000168} = 2980.$

$$\log \frac{vd}{\nu} = 3{\cdot}4742$$

The value of $\frac{mi}{v^2}$ from Stanton curve (top curve) corresponding to $\frac{vd}{\nu} = 2980$ is ${\cdot}0001925$

whence $i = {\cdot}0416$
and $h_f = {\cdot}0416\,(\mathrm{H} + 30)$.

Slippage.—$h_b = \left(\frac{\mathrm{H} + h_s}{v_m}\right) v_b$, taking $v_b = {\cdot}5$ ft. per sec.

The value of H actual lift is obtained from equation on p. 66, thus:

$$\left(\frac{\mathrm{H} + h_s}{v_m}\right) v_b + h_s + i\,(\mathrm{H} + h_s) + h_v + \mathrm{H} = h_e$$

$$\left(\frac{\mathrm{H} + 30}{3}\right) {\cdot}5 + 30 + {\cdot}0416\,(\mathrm{H} + 30) + {\cdot}14 + \mathrm{H} =$$

$$30 + 45{\cdot}15 = 75{\cdot}15,$$

$$\begin{aligned} &\text{from which } H = 32{\cdot}1 \text{ feet} \\ &\text{and } h = {\cdot}0416\,(62{\cdot}1) = 2{\cdot}58 \text{ ft.} \\ &\quad ,, \; h_b = \left(\frac{62{\cdot}1}{3}\right)\tfrac{1}{2} = 10{\cdot}3 \text{ ft.} \end{aligned}$$

The following diagrams (Figs. 40–43) for the pumping of sulphuric acid of 1·6 sp. gr. are given to show how the value of h_v, h_f and h_b, calculated according to the methods already suggested, affect the efficiency of the air lift for varying velocities and ratios of air to acid. A study of these diagrams will show that the empirical ratio of lift to submergence of 1 : 2 is not the most advantageous, and also that the usual habit of forcing air lifts is very wasteful. For a given submergence and lift the best efficiency lies round about 3 to 5 ft. mean velocity. Outside these figures the efficiency falls off rapidly. The labour involved in making the calculations is well repaid in air-lift design.

Pumping Hot Liquids in Air Lifts.—When used for pumping hot liquids the air lift is essentially a heat engine, so that instead of hot solutions being troublesome to pump, they enable the lift to attain what the pump engineer would term impossible efficiencies. As gases expand 1/273 of their volume for each rise of one degree Centigrade, it is obvious therefore that a given volume of air or gas admitted to the footpiece will expand more than the value, according to $PV = K$. How much this extra expansion amounts to is difficult to determine, as the transmission of heat from the liquid to the air through a surface film is almost as complex as the action of the lift itself.

It is not yet appreciated how easy the pumping of molten metals, alloys, mercury and fused salts is when measures are taken to use a gas for lifting which does not react with these metals. The author has raised fused sodium amide with a small lift, using pure hydrogen gas as aerating substance. In these cases the lift is of the simplest construction, merely consisting of small bore solid-drawn steel pipes. The gas pipe is usually ¼ in. in diameter and fitted inside the rising main, which is about ½ in. to ¾ in. Where the gas is expensive it is quite easy to design a closed circuit in conjunction with a compressor which circulates a given quantity of gas through the lift. A similar procedure is necessary in pumping strong acids which absorb moisture from the air.

The Granular Lift.—The ill report which the air lift has borne so long on account of its so-called low efficiency has prevented its special characteristics from being properly studied. The pumping of slimes, muds or granular materials of all kinds can be accomplished by special forms of ram, diaphragm, and centrifugal pumps, but when these fail then there is nothing for it but the bucket elevator and an extra page in the ledger for the maintenance account. The raising of tailings, the residue from the wet treatment of ore at the mines, is now accomplished by the air lift. The pulp consists of a mixture of one part of ore to about six parts of water.

At the Chino Copper Co., according to a report which appeared in *Engineering and Mining Journal*, November 19, 1921, the installing of the air lift in place of the bucket elevator reduced the costs about one-half. The following table, taken from the article just mentioned, speaks for itself.

TABLE 5.

BASED ON ACTUAL AVERAGE TONNAGE OF 5,000 DRY TONS DAILY FOR SIXTEEN MONTHS

Plant.	Tons Elevated 40 ft. in 16 months.	Actual Total Operating and Repair Cost, 16 months.	Actual Cost per Dry Ton Elevated.	Equivalent Cost per Year at full Capacity.	Estimated Operating and Repair Cost, 20 years.
Bucket Elevator . . .	2,459,300	$48,078·47	$0·01954	$85,585·20	$1,711,704
Air Lift . .	2,207,922	$21,798·90	$0·009873	$43,243·74	$864,874
Actual Saving	—	—	$0·009667	$42,341·46	$846,830

Other useful figures are given in the report which will repay study. Efficiencies measured from the energy in the air at foot piece varied from 50 to 58 per cent.

The design of lifts for raising granular bodies depends entirely on experience. The chief points, however, of the ordinary liquid lift apply, such as dividing the air into small bubbles, keeping down velocities, but in addition special arrangements have to be made to prevent the solid particles from packing and the air apertures from becoming choked. This little work cannot discuss details of the granular lift, but perhaps enough has been said to put makers of mechanical pumps on their mettle. Certain it is that when the best practice of these lifts becomes known many pump engineers will be surprised.

XI

MOMENTUM PUMPS

THE most notable example of this class of pump is the well-known ram pump (Fig. 44), which dates from the end of the eighteenth century. In its simplest form it consists of a

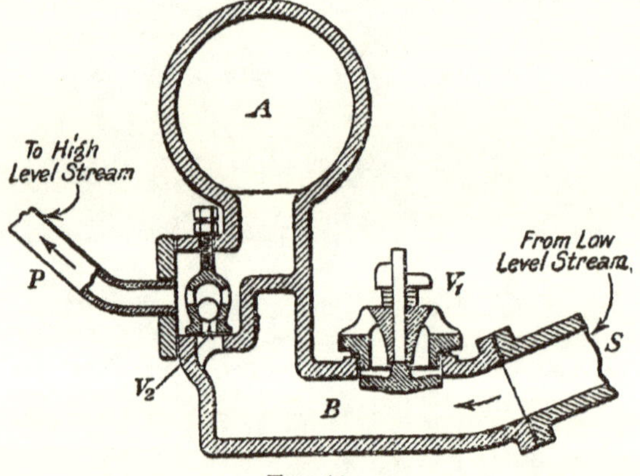

FIG. 44.

supply pipe S, a valve box B, a waste valve V_1, opening inwards, a discharge valve V_2, opening outwards, an air vessel A, and a discharge pipe P. Its action is as follows: A flow is set up in the pipe S on opening the waste valve V_1. The increase in momentum of the water closes this valve with consequent rapid increase of pressure in valve box B, and discharge through valve V_2 into the air vessel A. As soon as the momentum is absorbed in the valve box B, the delivery valve V_2 closes, and V_1 opens, and the whole cycle is repeated. The efficiency of the ram is a function of the delivery head, and the ratio 1/H, where 1 is the length of the supply pipe and H the effective supply head. Rankine gives the efficiency thus:—$1{\cdot}12 - {\cdot}2\sqrt{\frac{ha - H}{H}}$, where ha = delivery head.

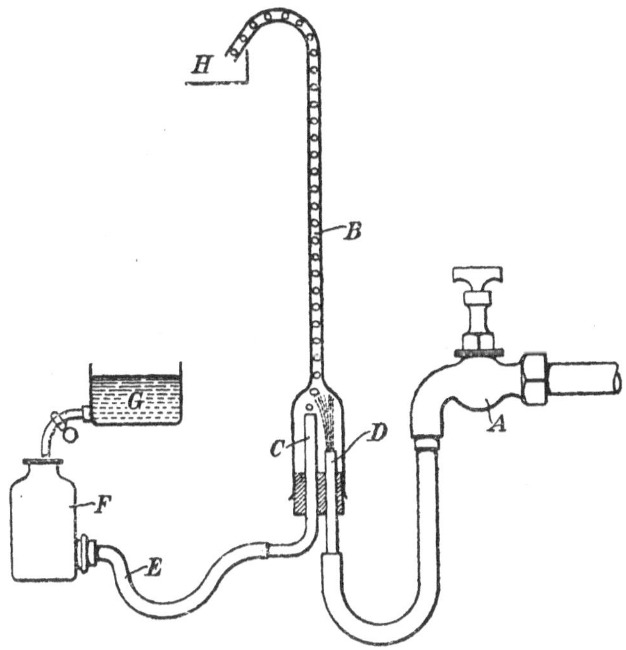

FIG. 45.—MERCURY MOMENTUM PUMP.

By inserting a rubber diaphragm in the valve box B between the two valves, it is possible to pump a small quantity of a liquid by the flow of a large quantity of another liquid. Use of the ram for air compressing was made by M. Sommellier in the construction of the Mont Cenis Tunnel.

A form of momentum pump exists in the form of a long vertical pipe, fitted at its lower end with a valve, and so arranged that it can be periodically jerked through a small range by an eccentric. The friction of the pipe acts as a retaining valve and causes the liquid to climb up the pipe.

An interesting form of this type of pump has been used by the author for raising small quantities of mercury by the flow of water from an ordinary service. The rising main B is fitted with an enlarged end to take a rubber bung for two glass tubes. The shorter tube D is connected to the water tap A, and the longer tube C is connected by a rubber play pipe E, with a mercury momentum jar F, fed from a container G. To operate, it is first necessary to permit the water to flow through pipe B, then jar F is gently lifted and jerked several times, till small globules of mercury issue from tube C. At this stage the jar F is placed on a stand so that the level of the mercury in it is the same as the tip of tube C. So long as the water flows, the mercury oscillates about a mean position, causing each globule of the metal issuing from C to act as a valve, and arresting for the time being the flow of the water.

Printed in Great Britain by Butler & Tanner, *Frome and London*

Publisher's Note

To facilitate the printing of this title, blank pages have been added. This does not influence the original retail price.

www.ingramcontent.com/pod-product-compliance
Lightning Source LLC
LaVergne TN
LVHW050933080826
845145LV00004B/1247

* 9 7 8 1 9 2 9 1 4 8 2 3 3 *